Iran

PAST AND PRESENT

IRAN

Past and Present

NINTH EDITION

From Monarchy to Islamic Republic

BY DONALD N. WILBER

PRINCETON UNIVERSITY PRESS

PRINCETON, NEW JERSEY

PREFACE

THE departure of Muhammad Reza Pahlavi from Iran on the 16th of January 1979, and the arrival in Tehran of the Ayatullah Khomeini on the 1st of February signaled the end of the Pahlavi dynasty and of the monarchial system of government which had prevailed in Iran for many centuries. Experienced observers of the Iranian scene were aware of spreading dissatisfaction within the country, but none of them foresaw that it could be welded into an irresistible force by the activity of one elderly man. From 1979 into 1981 a flood of books have traced the course of the revolution and its immediate aftermath. Serious efforts may be overshadowed by virulent attacks on the Pahlavi regime sponsored by radical groups in France, Germany, and Italy.

This ninth edition contains a concluding chapter entitled, From Monarchy to Islamic Republic. Too restricted in space to offer a detailed recital of events, it concentrates on a number of subjects. These include the activity of the Ayatullah Khomeini; an account of the groups that demonstrated against the regime; the new constitution of the Islamic Republic; the roles of President Bani Sadr and the leading clerics; the current problems of society, politics, and economy; and recent developments within the country.

An effort has been made to bring up-to-date, as nearly as possible, major statistical information. Beyond this effort, it is apparent that much of the material in other chapters is no longer in accord with the present situation. However, it is still too early to attempt to rewrite the sections on government, on natural resources, and on facilities, all of which may, in fact, be less altered in character in later months than now appears to be the case.

July 1981   DONALD N. WILBER

CONTENTS

IRAN'S HERITAGE

MODERN IRAN

Employees. The Prime Minister and the Council of Ministers. Foreign Affairs. Armed Forces and Other Organizations Concerned with National and Internal Security. National Budget. The Plan Organization. Indicators of Economic Growth. Justice and the Courts.

Minerals. Petroleum, Natural Gas, and Petrochemicals. Cultivation and Soils. Agricultural and Natural Resources. Water Supply and Irrigation. Fisheries. Farming Methods. Animals and Birds.

Industry. Foreign Investment. Rug Weaving. Industrial Labor. Housing. Foreign Trade. Banking. Public Finance. Electric Power. Telecommunications. Roads. Railways. Air Services. Ports and Navigation.

ILLUSTRATIONS

CREDITS FOR ILLUSTRATIONS

Iranian Information Center, 1, 3, 7, 8, 9, 10, 17, 19, 20, 21, 22
Arthur Upham Pope, 11
Donald N. Wilber, 2, 12, 13, 14, 15, 16, 18
Wilber Collections, 4, 5, 6
Wide World Photos, 21, 22

[ix]

IRAN'S HERITAGE

I. THE LAND

Place Name

IRAN and Persia: the two names have been used to designate the same country, but are not true synonyms. When the Aryan peoples migrated from their original territory, somewhere within Asia, to the upland plateau below the Caspian Sea, one of their tribal groups was the Iranian. The Iranian tribe called Parsa finally settled in a region of the plateau which they called Parsa. In time this regional name became Pars, and Fars, and hence the people of many other lands came to call the country Persia. In Sasanian times the official name of the empire of Iran was Iranshahr. Since 1935 when the Iranian government, for the sake of consistency, requested all foreign countries to use the official name of Iran, the correct designation has gained general usage. On the other hand the language of the country is Persian, *farsi* to the inhabitants, since it derived, over the centuries, from the language of ancient Parsa.

Location

Iran lies between the Caspian Sea and the Persian Gulf, and has common frontiers with Iraq, Turkey, Soviet Russia, Afghanistan and Pakistan. Along her perimeter dwell peoples of various languages or ethnic stocks whose area of settlement or of tribal movement overlaps the actual boundaries of the country.

Iran's geographical position made it the bridge for communication by land between Far Eastern Asia and the lands of the Mediterranean and Europe. Before the dawn of recorded history its mountain caves sheltered the hunters who were among the earliest people of the world to move down into the lower plains and to settle in villages, cultivate crops, and raise domestic animals. It also lay athwart the lines of movement of the early migrant tribes of central Asia, and became settled by many of these groups. Within historic times its rulers expanded their control far to the east and the west of the plateau and established the first great world empire.

For hundreds of years the main trade routes between the Far East and the West crossed northern Iran, and later on, when sea routes became of equal importance, additional highways led up from ports along the Persian Gulf to the principal commercial centers both within the country and beyond its frontiers.

The vital role of the overland trade routes across Iran was seriously limited by the construction of the Suez Canal, and the decline of her importance as a channel of trade heralded a period of political and military weakness. At the same time her strategic location made her a bone of contention between great powers whose interests were diametrically opposed. Her present frontiers, established during the nineteenth century, were the result of a series of wars in which she was unable to hold her own against more powerful neighbors.

Iran today covers an area of 628,000 square miles, approximately as large as that part of the United States which lies east of the Mississippi River, exclusive of New England—a much smaller area than at many times in her long existence. In general, her previous frontiers were much farther to the east than at present and fairly close to her present western ones, the greater expansion toward the east having been the result both of the less broken topography in that direction and of the strong linguistic and ethnic relations between the people of that area and the Iranians of the plateau. In spite of its present restricted size the country is still known as the Empire of Iran, and its ruler is the *Shahinshah*, or "King of Kings," a title first used in Iran over two thousand five hundred years ago.

Geology and Topography

The geological formations of the country have been fairly well studied, although not in a systematic fashion. More detailed studies of certain areas have been made in connection with the search for oil fields and for important mineral deposits.

The average traveler in Iran is made aware of its geological history by the lofty peaks, the ranks of jagged mountains springing abruptly from the fairly level plains, the vivid coloration of many formations, and the spectacular faulting or folding of the rocks. The traveler by air, who usually approaches Iran from the west or southwest, looks down on a series of mountain ranges which resemble the corrugated surface of a washboard. Each successive ridge is tilted from the vertical and higher than the one before until the general level of the high plateau is reached, but even there mountains rise on every side. From sufficient altitude the villages and tilled fields lose all identity, and the entire country seems to be barren and devoid of life.

Iran displays all systems from the Pre-Cambrian to the Quaternary, while from the beginning of the Palaeozoic era onwards fossils are rather abundant. In summary, after a turbulent start in the Pre-Cambrian there was a long lull until the Jurassic when the coal beds of the Alborz and the Kerman region were laid down. Movements began in central Iran in the Jurassic and reached a climax in the Upper Cretaceous, when some folding took place with volcanic activity, northeast of the Zagros mountains. Volcanoes remained active throughout much of the Eocene. With some breakdown due to faulting, folding increased to a maximum in the Pliocene or Mio-Pliocene.

The Zagros and Alborz ranges, formed by the Mio-Pliocene, attain altitudes of over 11,000 feet. As mountains go they are rather young, as is shown by their sharp, broken profiles. The general configuration of these ranges and of the Iranian plateau seems to have been the result of prolonged pressure against the area of Iran from a Russian mass on the north and an African mass on the south. The fact that the southward pressure was the stronger is indicated by the steeper slopes of the Alborz as compared to the softer folds of the Zagros Range. The building of the mountain systems was of course complicated by vertical movements and by extensive faulting.

In several regions of the country the prominent cones of formerly active volcanoes are a dominant feature of the landscape. The principal volcanic peaks are Demavand, the highest peak in Iran, which figures in many ancient tales of heroes and demons, rising to 18,600 feet, in the north; Savalan, at

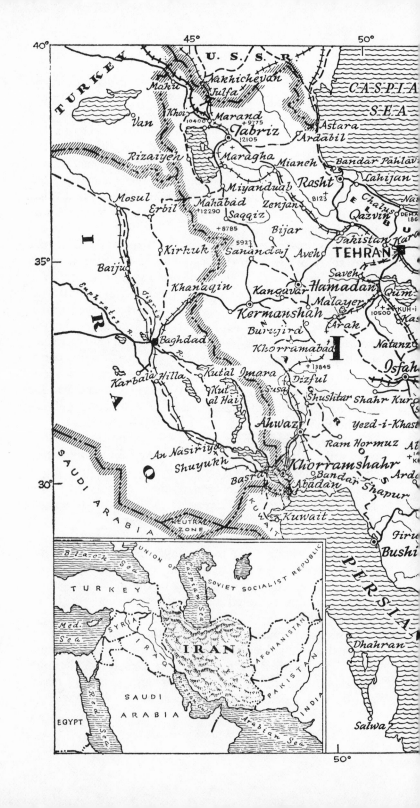

14,000 feet, Sahand at 12,138 feet, and Ararat (lesser Ararat is within Iran, but greater Ararat lies just over the frontier) in the northwest; and Bazman and Kuh-i-Taftan at 13,262 feet, in the extreme southeast. Two of these cones still show some traces of activity: all climbers who have made the ascent of Demavand have noted the presence of sulphur gases, and Kuh-i-Taftan occasionally spews forth gas and mud.

Oil was trapped and stored in Asmari limestone of the Oliocene and Lower Miocene, with this limestone formation stretching for nearly five hundred miles. Some 300 salt plugs, ranging from a half mile to four miles in diameter, are found in the area south of Tehran and strewn along the shore of the Persian Gulf.

Iran may be described in general terms as a high plateau some 4,000 feet above sea level, strewn with mountains. Specifically, there are four main topographical areas, each distinctive in character and extending beyond the frontiers of Iran:

1. The great Zagros and Alborz Ranges, stamping a huge V upon the surface of the country. The apex of the V forms in northwestern Iran and extends beyond into Turkey and the Russian Caucasus. The southern arm of the letter is represented by the Zagros Range, which runs southeast and roughly parallels the frontier of Iraq, while the upper arm, the Alborz Range, looms like a great wall across the north of the country, breaking down into other ranges which run on into Afghanistan and Turkestan.

2. The area within the V begins as the high plateau with its own secondary mountain ranges and gradually levels off to become the empty deserts which continue into southern Afghanistan and Pakistan.

3. The region of Khuzistan, below the lower arm of the V, is a continuation of the low-lying plain of Iraq.

4. The Caspian Sea coast above the upper arm of the V is below sea level and forms a separate climatic zone.

The fact that each of these topographical formations extends beyond the frontiers of Iran does not mean that the country is easy of access, for its present boundaries are guarded by formidable natural barriers. The entire western frontier and the country inland from the Persian Gulf is protected by lofty ramparts of rock, where steep passes lead from sea level to a height of over 7,000 feet and down again to the plateau. Equally forbidding passes isolate the Caspian littoral from the rest of Iran, and along the northeastern and eastern frontiers the approaches are either through hilly country or across vast spaces of empty desert.

Drainage

There are four principal drainage basins, roughly corresponding to the topographical zones: The Caspian, the Lake Rezaieh, the Persian Gulf, and the great desert basins. The Persian Gulf basin is fed by three separate systems: the smaller streams of northwest Iran which eventually find their way into the Tigris River; the Karun and its tributaries, which empty into the head of the Persian Gulf; and the countless streams which indent the thousand-mile-long coast line of the Persian Gulf.

Most of the rivers and streams of Iran flow not into one of the three large bodies of water, but into the vast interior deserts where there are three subordinate drainage areas, separated from one another by lines of hills but having the same effect as that of a single basin. This feature of interior drainage has an important relation to the economic life of Iran.

The great majority of the inhabitants of Iran live along the lines of the V formed by the main mountain ranges. The mountains run in parallel files, enclosing long, narrow valleys walled at each end by mountainous cross barriers. The general pattern of mountains and valleys may be compared to a number of ladders laid down roughly parallel to each other, the uprights representing the lines of the mountains and the rungs the barriers across the ends of each valley. An average valley may be eight miles in width and from 25 to 40 miles long, flat bottomed, with rims rising directly and abruptly into the mountains above. Villages are more closely clustered along the rims than along the center line where may be found one of the highways which even in modern times have penetrated relatively few of the thousands of mountain valleys. Nomads spend the summers in the higher altitudes where the heavy snows and extreme cold of winter would make village life impossible, while farming communities abound in the valleys where the level ground is more suitable for cultivation. For centuries the farmers have led isolated, self-sufficient lives, and the barriers which separate them from the outer world have contributed to historical continuity, preservation of racial stocks, and comparative safety and independence in periods of foreign invasion.

Rivers

The more important rivers which flow into the Caspian are, from west to east: the Aras (Araxes), the Sefid and its long tributary the Qizil Uzun, the Chalus, the Haraz, the Lar, the Gorgan, and the Atrek. None of them carry a great volume of water except in the spring. Their lower courses serve as spawning grounds for the sturgeon of the Caspian, productive of fine caviar, and these streams offer a reservoir of water power.

The largest river flowing into the Persian Gulf is the Karun, whose lower course is the confluence of the upper Karun and the Diz River. The Karun is navigable by small steamers as far as Ahwaz, some 70 miles from its mouth, and smaller boats can proceed even farther above the Ahwaz rapids. At Ahwaz the river is about one hundred yards wide, and is spanned by highway bridges and a railroad bridge.

Rivers which empty into the interior desert basins of the country are the Zayandeh, which flows past Isfahan; the Jajirud, Karej, and Kand, near Tehran; the Qarasu, near Hamadan; the Hableh, east of Kashan; the Qum, flowing by the town of the same name; and the Kur, which runs near ancient Persepolis. Far to the east is the Helmand River, whose dammed-up waters, although nearly all of its course is within Afghanistan, are led through channels to irrigate a part of the province of Sistan.

These rivers are the principal perennial streams; most of the others run dry in the middle of the summer. Typical of the larger streams in this category is the Zayandeh, whose waters nourish Isfahan and the score of adjacent farming communities. In the early spring this river is more than one hundred yards wide at Isfahan and nearly six feet deep. Later in the spring the communities upstream from Isfahan draw off its water to their fields, or flood paddies for cultivating rice, and as a result, in midsummer the river opposite Isfahan is just adequate for the needs of the area.

In recent years Iran has constructed a number of great dams which provide water for irrigation, create electric power, and, in some cases, water for urban consumption.

Seas and Lakes

The Caspian Sea, the largest landlocked body of water in the world, lies some 85 feet below sea level, is comparatively shallow, and for several centuries has been slowly shrinking in size. Its salt content is considerably less than that of the oceans, and though it abounds with fish its shelving coasts do not offer any good natural harbors, and sudden and violent storms make it dangerous for small boats.

Lake Rezaieh, formerly called Lake Urmiya, is about 80 miles long and 35 miles wide, or approximately the same size as the Great Salt Lake of Utah, and averages from five to six yards in depth with a maximum depth of eleven yards. Only a few minor streams flow into this landlocked body of water, which has resulted in a steady shrinking in the size of the lake and so high a concentration of salt that no fish can live in it. Its saline content, made up of salts and sulphates of magnesium, calcium, sodium, and potassium, is as high as 23 percent. Most maps show a large island of Shahi well out toward the center of the lake, but within recent years the water has receded so far that during the summer months it is now possible to walk directly from the mainland to the shores of the "island."

Along the frontier between Iran and Afghanistan are several marshy lakes which expand and contract according to the season of the year. The largest of these, the Hamun-i-Sabari, is alive with wild fowl. Real fresh water lakes are exceedingly rare in Iran; there are probably not more than ten in all the country, all of them brackish and most not much bigger than ponds.

Deserts

The vast desert regions of Iran stretch across the plateau from the northwest, close to Tehran and Qum, for a distance of nearly 800 miles to the southeast and beyond the frontier. Approximately one-sixth of the total area of Iran is barren desert.

The two largest desert areas are known as the Dasht-i-Lut and the Dasht-i-Kavir, *dasht* meaning "plain," *lut* meaning "naked" or "barren," and *kavir* meaning "salt desert." Third in size of these deserts is the Jaz Murian. Some sections contain salt lakes which each year are swollen by the spring torrents from the interior rivers and in summer disappear beneath a hard salt crust. Other areas are stony wastes, wide stretches of saline soil or deserts heaped with sand dunes, and nearly all are crisscrossed by ranges of hills. At a few widely separated points copious springs flowing from the naked earth have created fertile oases, such as the charming village of Tabas, set in the midst of flowering gardens, orchards, and groves of date palms.

It is often said that the Dasht-i-Lut and the Dasht-i-Kavir are impassable except by the single road which runs from Yazd to Firdaus, but within recent years heavy trucks and other vehicles have traveled over long stretches of these deserts. Ruins of caravanserais and of villages found there are evidence that not too long ago the deserts were kinder to human existence. For example, in fairly recent times there was a direct north-south caravan road from Simnan to Kerman: a section of its *sang farash*, or "stone carpet," some 18 miles long and up to 40 feet wide still exists. The labor involved in conveying these stones great distances must have been enormous. However, it is true that disaster dogged the caravans; many fell victim to exhaustion of water supplies in summer or to winter's sudden rains which could turn the surface into a slippery morass and always chance straying from the trail might mean, for men and camels, breaking through the salt crust to become hopelessly trapped in the mire beneath.

[12]

These deserts contain extensive mineral deposits—chlorides, sulphates and carbonates—and it is only a matter of time until they are exploited.

Climate

Rainfall in Iran, the result of atmospheric depressions moving eastward from the region of the Mediterranean Sea, is largely confined to the winter months from November to early April. Summer thunderstorms, so familiar to us, are quite rare, and the occasional rains in midsummer are limited to light showers.

Over most of the Iranian plateau the total annual rainfall is less than twelve inches, or about that of the state of Nevada. The extensive desert regions and the southeastern corner of the country receive less than five inches of rain a year. The northwest corner of the country often benefits by from 15 to 35 inches, and it is there that dry farming, the growing of unirrigated crops, is most widely carried on. However, even in areas receiving the minimum amount of precipitation the farmers will sow a limited amount of winter wheat with the knowledge that if the winter brings three or four good rains there will be a crop worth harvesting. Torrential downpours sometimes occur, and when this happens flash floods sweep down from the mountains to wash out roads, damage villages, and destroy crops.

The Caspian littoral presents quite a different picture, for there the annual rainfall is from forty to sixty inches and rain falls throughout the year, creating the extensive marshes and the dank jungle foliage which breed the fevers so common in this region.

Rain in the valleys means snow on the high mountains, and many of the higher peaks are crowned with snow until late in the summer. On the valley plains heavy snowfalls are comparatively rare, and in such towns as Tehran, Isfahan, and Kermanshah the snow usually melts away in a few days. Snow does not normally fall south of a line connecting Andimeshk, Shiraz, Yazd, and Qaen, and seldom along the Caspian coast. But the high passes of the main highways may be blocked for days or even weeks at a time in the late winter, and the Chalus road from Tehran to the Caspian Sea is usually closed for three months in the year. The amount of snow in the high mountains also controls the spring migrations of the nomadic tribes, whose flocks often cannot move through certain passes until May. The Iranian plateau enjoys fairly mild winters and hot summers. For example, at Tehran there may be a period of several weeks in midsummer when the temperature reaches 100 degrees day after day, but the nights are relatively fresh and cool. The humidity in summer is only about 20 percent. In winter the temperature rarely reaches zero and usually is above the freezing mark in the middle of the day. The regions of Khorasan and Azerbaijan are cooler in summer and much colder in winter.

Much less tolerable conditions prevail at the head of the Persian Gulf, along the shores of the Gulf, and in the desert regions of the southeast. It is a question whether temperatures are higher at Abadan, Khorramshahr, and Ahwaz, or in the deserts. Both sections have recorded readings of 130 degrees, but the area at the head of the Persian Gulf is also plagued by a high humidity. The Caspian littoral ranges from warm to hot throughout the year, and in January, the height of the rainy season, the humidity averages 90 percent while in July it is about 75 percent.

In Iran the change from one season to the next is fairly abrupt. By March 21, the first day of the Iranian year, the fruits trees are in full bud and fresh green wheat covers the fields, and later, while the orchards are in bloom, wild flowers carpet the stony hills. Summer heat burns and kills the flowers. Autumn is not marked by a display of bright colors and the soft haze of Indian summer; instead there is a rapid transition from summer to winter.

Clear days are the rule in Iran, for the skies are cloudless for more than half the days of each year and only on about fifty are they stormy and overcast. Strong winds blow at certain periods and often fill the air with a haze of dust. In southeast Iran summer winds blow steadily over a period of weeks with a velocity that exceeds sixty miles an hour, and their force is utilized by windmills to grind flour from the local wheat harvest.

The cold of winter is more dreaded by the Persians than the heat of summer, for travel and transport may be forced to a halt and the farmer must conserve his limited supply of food. Even more serious is the fact that the cold cannot be countered by a system of heating. All houses of any size are turned toward the south so that the welcome winter sun, which can be very warm, will shine directly into the main room of the house. All types of fuel are scarce and expensive, and in the villages there is no way of heating the entire house. Families keep warm by using a *kursi*, which is a pan or brazier containing slow burning charcoal. The pan is placed on the floor with a circular table or frame about two feet above it, and on top of the table all the family quilts are draped to form a circle some three or four yards across. Members of the family may recline during the day, and sleep at night, under the quilts, with their bodies radiating like spokes from the hub of the charcoal pan and only their heads exposed to the cold air of the room.

II. HISTORY

Prehistoric Iran

Ⓜ AN's progress from the remote time of his emergence as a species down to the present day may be shown on a scale fifty inches long, of which each inch represents 10,000 years. According to such a scale, man became a farmer at the 49¼ inch point, learned to write at 49½ inches, and at 49¾ inches culture and civilization, such as that of the Achaemenid empire, was in existence. Most of the material which follows in this chapter will deal with the final quarter inch on the scale and with the long struggle of man against himself, against his fellow men, and against nature. But some mention must be made of the fascinating prehistoric period, the time before written records were made.

The Palaeolithic or Stone Age men of prehistoric times were hunters. They fashioned rough weapons and implements, pursued the game upon which their lives depended, and took shelter in caves. The population of any area was small, for it bore a direct ratio to the number of wild animals in the region and starvation was ever-threatening. Groups probably moved within fairly well-defined areas, but the pressure of hunger led to wider movement and there seems to have been considerable intercourse over wide regions at very early periods.

The earliest evidence of man found thus far in Iran comes from the Middle Palaeolithic period. In Shanidar, a large cave in the Zagros range in northern Iraq, just across the Iranian frontier, six skeletons were found. According to the Carbon-14 method of dating, the material is from 60,000 to 46,000 ± 1500 B.P. (Before the Present time), and the finder said that these people were classic Neanderthaloids of the Mousterian culture period. In 1964 C.B.M. McBurney excavated a cave near the northeastern corner of the Caspian Sea and found material with Carbon-14 dates of 65,000 to 40,000 B.P. Bones found in the cave indicated that these people hunted the rhinoceros, wild horses, and bears. Flints found by Carleton S. Coon in caves at Bisutun may well date from these same ancient times.

In the Upper Palaeolithic period, 20,000 to 15,000 years ago, the so-called Aurignacian man, a Post-Neanderthal type, probably was active on the plateau. There is evidence of his activity in the Middle Palaeolithic and Mesolithic periods. In 1950 Carleton S. Coon excavated the caves of Hotu and Kamarband along the Caspian Sea. The Carbon-14 dating of material from Kamarband was 11,480 ± 550 B.P. In the caves were found flint cores, flakes and blades, and bones which indicated that the people were hunters of seals and gazelles. Near the end of the occupation period of the caves they were producing Neolithic painted pottery. The caves of Hazar Mard, Zarzai, and the upper levels of Shanidar, yielded material of this same general period.

Three vital discoveries led to a marked change in mode of life and probably in social organization. These were the introduction of cultivation, the use of metals, and community life. It is probable that primitive people had long been gathering wild rice and wheat and roots, but the decisive step was taken when cereals gathered in the fall were first stored and sowed in the spring. This control over the means of subsistence resulted in group movements from the higher mountains down to the flat-bottomed valleys where level surfaces lay ready for cultivation, a major change for which the topography created both the means and the stimulus. The wild grains such as wild emmer, the ancestor of wheat, which still grows in Iran, were native not to the basins of the great rivers of the Middle East but to the mountain slopes, and the challenge of and response to man's natural surroundings were stronger in the cool hills than on the hot plains. Planted fields soon surrounded compact settlements and animals were domesticated within them, and the new type of community life led to changes in the social relations between individuals and between groups. Until quite recent years it was believed that this transition from hunting and food gathering to food production in settled communities took place about 6000 B.C. Now, however, surveys of hundreds of prehistoric sites and the excavation of a number of mounds have served to push back this time of transition to between 15,000 and 8,000 years ago. The mound of Sarab yielded clay figurines, pottery, and stone, flint and obsidian tools, and probably dates from 9,000 or 8,000 years ago. Ganj Dareh, located, as is Sarab, in the general vicinity of Kermanshah, was excavated after 1965 and its Carbon-14 dates are 8500 ± 150 B.P. Its houses of mud brick probably were the winter quarters of a small community which summered in the hills. Tepe Yahya, a huge circular mound between Kerman and Bandar 'Abbas, was continuously occupied from about 4500 B.C. until A.D. 400. Similarly, Yanik Tepe, in the northwest, has revealed a cultural sequence of from about 6000 B.C. down to Islamic times.

On the plains of southwestern Iran a number of early sites have been excavated. Tepe 'Ali Kosh dates back to 7500 B.C., and Choga Mish and Choga Sefid are nearly as old.

By the end of the Neolithic period, that of polished stone implements, large villages were scattered over the level plains. Each settlement was packed with closely crowded houses and there was no orderly system of streets and lanes. After 3000 B.C. bronze weapons, implements, and other objects must have been made in vast quantities, judging by the mass of bronzes recovered in archaeological excavations. The date at which copper was first smelted from ore in Iran has recently been pushed back to the fifth millennium B.C., but that at which bronze was first made from copper and tin is still uncertain.

Scores of these village sites have been excavated. Broadly speaking, they are to be distinguished from the sites already mentioned because of their larger settlements and the quantity, quality, and variety of the artifacts found. Only a few of these sites can be mentioned.

Tepe Sialk, near Kashan, was the first such site to be excavated, and is probably the one best known to scholars and laymen alike because of the exquisite hand-made and wheel-turned pottery from several occupation levels. Notable were vessels with very long spouts which were decorated with the figures of naturalistic animals.

Tepe Giyan, like Tepe Sialk, was excavated in the 1930s

by Roman Ghirshman. Located near Nihavand in western Iran, it was occupied from the fifth millennium down to about 1000 B.C., and it too brought to light striking pottery of the third and second millenniums.

Near Persepolis is Tall-i-Bakun, excavated in the 1930s and again some twenty years later. This site may go back to 4500 B.C., and its distinctive pottery includes bowls decorated with the elaborately curved horns of the wild mountain sheep.

Tepe Hissar, near Damghan, was also excavated in the 1930s, and its levels range from about 3500 to 1600 B.C. The pottery from its earliest period has relations with that found at Tepe Sialk.

The sites of early settlements appear as conspicuous mounds, some nearly one hundred feet in height, of which there are many thousand in Iran. The original settlements were founded on the level plain or upon a slight natural rise. After heavy rains the mud houses collapsed and new ones were built on the smoothed debris, and rubbish was thrown out in the narrow lanes so that their level was constantly being built up. Thus, in the course of centuries the settlements rose higher and higher above the plain. Occupation of a site would come to an end when the area at the top of the artificial mound became too small or when the population was wiped out by war or natural calamities. Over the deserted site wind blew sand and earth, and grass grew up to cover all traces of human habitation; but heavy rains still wash pieces of pottery, seals, and other small objects down to the base of the mound to reveal the presence and approximate age of an ancient settlement.

Excavation begins at a selected site by cutting a vertical trench through a part of the mound. If the finds of buildings and artifacts are promising, either a limited section or the entire mound is slowly cleared. The technique of digging resembles the cutting away of a series of horizontal slices. Within each slice the vertical position and location plan of every building and object is carefully recorded, so that it is possible to detect marked changes in construction and in types of pottery and other objects. Distinctive horizontal levels can be given a letter or number designation, and similarity of levels between different sites used to establish common cultural traditions or to indicate influences from one region toward another. This system of comparative levels also permits the estimation of dates for periods antedating written records. The objects found at an ancient site include tools, implements for cultivation and weaving, seals, cult figurines, children's toys, and pottery. Pottery, found in large quantities at each site, yields definite information, and the designs offer a fertile field for speculation. The earliest pottery was handmade, but later the potter's wheel came into use. There is great variety in the size and shape of the vessels dug up, and many of them, so thin and fragile that they could not have been suitable for daily use, have been found in graves, suggesting that they were made to provide for the needs of the deceased in and beyond the grave. Vessels painted in red or black are decorated with geometrical patterns, lively animal figures, and, much more rarely, figures of human beings. The animals common to the region are depicted with amazing skill, and it is especially interesting to note that the village artists made sometimes very naturalistic and sometimes highly stylized pictures of the same beasts. They are always shown in profile, and represent the persistent memory image of the creator rather than a direct copy of the animal's form.

It seems possible that all the geometrical patterns used on the pottery had a symbolical meaning for the artists and the people of the time, representing man's halting efforts to grasp the significance of the world, his interest in his surroundings, and his awe of the forces of nature. Dependent as he was upon favorable weather for the crops upon which he subsisted, many of his designs symbolized natural forces involved in the weather or were a transference of the symbol to similar animal form. For example, the moon, associated with rain, is symbolized by a crescent, and the crescent is then represented by the curving horns of the ibex. Others represented the sun, falling rain, a great pool, and a flourishing tree upon a mountain top. The symbols also signified evolving religious beliefs and myths which became so firmly entrenched in the minds of the people that they were handed down to be recorded finally in writing.

Sites of later dates will be mentioned in following pages; they are even more numerous than those of the very early times. In any one season, usually in the summer and fall, a score of important sites are being excavated or cleared by teams of Persian, American, British, French, German, Italian, and Japanese archaeologists and anthropologists from museums and universities. Permits for such work are obtained from the Archaeological Service of Iran, and scholars are attracted to the country not only because of the abundance of sites, but because the law provides for the equal division of finds between the excavators and the Archaeological Service, an agency of the Ministry of Culture and Arts. Unique pieces, such as, for example, gold and silver bowls, figurines, and jewelry, are not included in the divisions, and remain in Iran where they are on display in the splendid Archaeological Museum in Tehran.

There is also a regrettable amount of illicit digging at ancient sites. According to law no antiquities may be exported from Iran, but many very important objects obtained illegally are smuggled out of the country to come into the hands of museums and private collectors. The foreign archaeologists are opposed to this traffic, but museums in Europe and the United States continue to purchase objects of unknown provenances. There is, in addition, a heavy traffic in forgeries of antiquities, some produced in Iran and others abroad, and collectors may be taken in by the best of these forgeries.

Mention has been made of the Carbon-14 method of dating: this involves a naturally radioactive carbon isotope with a half-life of 5,700 years which is present in very small amounts in wood and other materials containing carbon. Specimens from ancient sites are examined in a laboratory and a dating obtained. Other methods for examining and dating artifacts are also used. Chemical analysis is applied to coloring matters, glass and glazes and metals, and radiographs are made of metal objects. Archaemagnetism identifies the magnetic fields in baked clay, such as bricks and pottery, and establishes dates. By neutron activation analysis the provenance of pottery and pottery glazes may be determined. Fission-track dating concerns zircon crystals which contain uranium concentrates and which are found in ancient pottery: this method gives results comparable to the Carbon-14 dating system. Finally, the electron-probe microanalyser is used to test glazes and metal: it can identify the relative amounts of various metals in a single object.

These highly sophisticated methods of examining and dating antique pieces are of great value not only to the archaeologist; they also make it increasingly difficult for modern forgeries to escape detection.

Pre-Achaemenid Iran

History is distinguished from prehistory by the existence of legible records of man's activity, and as history begins the area of western Iran is dominated by military and cultural influences from Mesopotamia to the west. Although the highlands had led the way in the development of farming and in community life, the people of the Mesopotamian plains—the land of the two rivers—came into the ascendancy with the invention of writing, the codification of law, and the establishment of monumental architecture.

While Mesopotamia was highly literate, the peoples of Iran remained in the Protoliterate stage, that is, just on the verge of producing their own written records. Who these aboriginal inhabitants of the plateau were and what were their ethnic relations to each other and to other groups may never be known. One suggested name for them is the Caspians, on the assumption that they had moved south from the region around the Caspian Sea, while a more neutral designation is that of the Asianics. Only after written records are available can local groups be reliably identified, either by the names they used or what they were called by more literate neighbors. And the records reveal that each such group was ruled by a line of kings.

Most important were the Elamites who established a high culture at Susa, their capital on the plain of the southwestern corner of the plateau—the present province of Khuzistan—and also in Anshan, across the Zagros range to the southeast. By 3000 B.C. they were using a Proto-Elamite script, a semi-pictographic writing, which remains unread. Later they developed their language by using the cuneiform script, and many of the legible clay tablets deal with commercial transactions. Curiously enough, this language is neither of the Semitic nor the Indo-European group. Often at odds with successive dynasties in Mesopotamia, about 2000 B.C. the Elamites destroyed the ancient city of Ur.

Elam was most prosperous in the thirteenth century. About 1250 B.C. King Untash Gal erected a ziggurat at Choga Zanbil, some 45 miles south of Susa. It was dedicated to the god Inshushinak: of sun-dried bricks lined with green and blue glazed bricks, it was over 300 feet square at its base and towered high over the city of Dur Untashi.

The great mound of Susa was identified over a century ago and members of the French Archaeological Mission in Iran have been excavating the site year after year for half that time. About 640 B.C. the Assyrians swept into Elam, set Susa ablaze, and sent the kingdom into a period of decline.

North of the Elamites were the Ellipi, Kassi, Lullubi, Guti, Mannai, and Urartu.

The kingdom of the Ellipi lay just north of Elam in the upper Kharka river valley and on to the plain of Kermanshah. It was probably centered at Harhar, depicted as a walled town on a river in a relief in the palace of Sargon at Khorsabad. Long vassals of the Assyrians, they are mentioned in an Assyrian record of about 900 B.C.

The Kassi, or Kassites, occupied the mountain valleys now known as Luristan, and possibly as far north as Hamadan. They are mentioned in an Elamite inscription of the twenty-fourth century. About 1750 B.C. they seized Babylonia and remained in control of that region for more than 500 years, and were finally conquered by the Elamites in the twelfth century. From their area have come the renowned Luristan bronzes—weapons, tools, figurines of gods, men, and animals, votive objects, horse trappings, jewelry, plaques, and household utensils—made by the lost wax process of casting. Especially striking are the animal forms which decorate so many of the pieces. The first found their way to the dealers of antiquities in Tehran in 1928, and in the years following thousands appeared as the local people conducted rather systematic searches for the graves in which they were placed with the bodies. The dates at which they were made was long hotly disputed by scholars, but in very recent years controlled excavations have made it certain that they were produced from 2600 B.C. to 600 B.C. Hence, it seems reasonable to assign these bronzes to the Kassites.

The Lullubi were settled along the ancient highway across the Zagros range, between Kermanshah and the plains of Mesopotamia. At the village of Sarpol is a rock-cut relief depicting King Anubanini trampling on naked prisoners and confronted by the goddess Inanna (Ishtar) who holds out the symbol of royalty. It was probably executed between the twenty-first and the nineteenth centuries. Three rather similar reliefs are nearby. In the eighth century the Lullubi joined the Guti in an unsuccessful attack on Assyria.

The Guti were to the north of the Lullubi in the mountains of Kurdistan. They invaded Babylonia as early as 2200 B.C.

The Mannai, or Manneans, were also in Kurdistan, principally in the region to the southeast of Lake Rezaieh. They are first mentioned in an Assyrian inscription of between 858 and 820 B.C. Two sites may be identified as settlements of the Mannai. The excavated mound known as Hasanlu—its early name is not known—revealed extensive structures and many important artifacts, including a gold bowl, eight inches high and eight in diameter, engraved with a number of mythological scenes in relief. It is believed to have been made between the thirteenth and the eleventh centuries.

At the village of Ziwiye a chance find in 1947 brought to light ivory plaques and figurines, pectorals, bracelets, plaques, dagger sheaths of gold, objects of silver and bronze, and glazed pottery. Many of these objects are in the museum at Tehran, and the others are in museums abroad. Later investigation indicated that the material was related to a citadel built on the crest of a hill, and that they date from the eighth and seventh centuries B.C.

The kingdom of Urartu, which rivaled Assyria in its power, was centered around Lake Van. The earliest known inscriptions of this people date from the ninth century. Current sur-

1. Pasargadae. The tomb of Cyrus, erected about 529 B.C.

2. Gorgan. The tomb-tower called Gunbad-i-Qabus, completed in 1007 A.D.

3. Persepolis. One of the monumental sculptured stairways leading to the *apadana*, or audience hall

veys and excavations reveal that they had numerous fortified settlements in the regions to the west and north of Lake Rezaieh.

At a number of times over many centuries conditions in Central Asia, such as pressure from aggressive neighbors or the overpopulation of grazing areas, resulted in mass migrations towards the west. The earliest such people to cross the Oxus River and move on to the Iranian plateau may have been the Aryans in the seventeenth century B.C. These Aryans—with Aryan meaning noble, or lord—were a nomadic people speaking an Indo-European language and were engaged in raising horses and cattle, and also had some experience in agriculture. Successive waves of these migrations continued for centuries. Between 1500 and 1200 B.C. numbers of these people moved down into the Indian subcontinent, while others continued to the plateau. Some of the latter groups stayed in eastern Iran, while others pushed on towards the Zagros range. These included the Iranians, comprising the Mada, or Amadai (Medes), and the Parsua, or Parsa (Persians), and the Scythians. Each of these groups was a confederation of tribes, with each tribe further subdivided into clans and families. Even today the tribes of Iran and Afghanistan display a similar structure.

The Iranians may have reached the plain to the east of Hamadan after 1200 in sufficient numbers to take over the area from its inhabitants. Before 900 they had pierced the difficult Zagros range and spread to the north and the south. While these dates are somewhat speculative, the fact that excavations of Iron Age I Period (c. 1300/1250 to 1000 B.C.) sites in this general region reveal that a plain, gray pottery was introduced at this time, and this pottery has been associated with these new arrivals.

The Medes established themselves over an extensive area east of the Zagros range—an area roughly bounded by the modern cities of Tehran, Tabriz, Hamadan, Isfahan, and Tehran—as well as in smaller numbers west of this range. They are mentioned in an Assyrian inscription of 836 B.C., and suffered attacks from the Assyrians and paid tribute to that kingdom. The Assyrian records also indicate that they were in direct contact with the Ellipi and the Manneans.

The Persians are first mentioned in an Assyrian inscription of 844 B.C., and later Assyrian records indicate that after the beginning of the eighth century they were allied with the Ellipi, the Elamites, and the people of Anshan. Both the earlier groups and the Iranians had mutual interests in trying to hold off Assyrian campaigns against the plateau. Apparently the Persians were in three separate regions: to the south of Lake Rezaieh, on the northern borders of Elam, and in what was to be their final homeland, the area of Parsa.

The Scythians were probably in the northern Zagros range, and had the reputation among the Assyrians of operating as marauding bands of warriors.

The waning power of the states of Urartu and Manna favored Median expansion. The names of Median leaders, such as Daiaukku and Kashtaritu, appear in Assyrian records, as does that of Uvakhshtra, founder of the extensive Median kingdom. Now, the Greek historians knew Uvakhshtra as Cyaxares, and in the following paragraphs the correct names of Medes and Persians are followed by the more familiar Greek versions.

In 614 B.C. Uvakhshtra invaded Assyria and took Assur, and two years later returned, with his Scythian and Babylonian allies, to capture and raze great Nineveh. After a pause to divide the booty, Uvakhshtra campaigned as far to the west as Lydia. In 585 he returned to absorb the state of Urartu, and to extend his realm as far to the east as the eastern end of the Caspian Sea. Uvakhshtra was succeeded by his son Ishtumegu (Astyages).

Meanwhile the Persians were led by Hakhamanish (Achaemenes), the head of the family of this name and the eponymous ancestor of the Achaemenid dynasty. Hakhamanish was followed by his son Chishpish (Teispes) who held power from about 675 to 640, and it was at the latter date that the Median leader Kashtaritu compelled the Persians to become vassals of the Medes. Nevertheless, Chishpish was able to expand his territory to include Anshan and Parsa. On his death one son, Ariyaramna (Ariarmnes) received Anshan and Parsa, and the other, Kurush (Cyrus I) took the Elamite region. In a gold tablet Ariyaramna called himself, "great king, king of kings, king of the land of Parsa."

The line of Kurush prospered as his son and successor Kambujiya (Cambyses I), in power from 600 until 559, was permitted to marry a daughter of his suzerain Ishtumegu. This marriage produced another Kurush (Cyrus II, known to history as Cyrus the Great), who united the Persian tribes. In 550 he led them against Ishtumegu and defeated the Median king in battle. A close union of the Medes and Persians followed, and an army drawn from these resurgent tribal groups was soon engaged in a series of successful campaigns which resulted in the establishment of the first world empire.

Cyrus drove on into Asia Minor where he defeated the Lydian King Croesus of fabled wealth, and by 546 B.C. controlled Armenia, Asia Minor, and the Greek colonies along the Mediterranean shore. Returning to Hagmatana, modern Hamadan, called Ecbatana by the Greeks, the commercial center of his Achaemenid empire, he then led an army east to conquer such regions as Parthia, Chorasmia, and Bactria. In 539 B.C. great Babylon was besieged and finally taken through the stratagem of diverting the Euphrates River, and some of the Jews held there in exile were sent back to Palestine. Cyrus, who died in 529 B.C., was not only a world conqueror and effective organizer, but the first to display that spirit of tolerance which is typical of the Iranian character.

His son Kambujiya (Cambyses II) conquered Egypt, and later fell insane and died by his own hand near Ecbatana during a revolt led by a *magush* or priest named Gaumata. Gaumata, who falsely claimed to be a brother of Cambyses, held the throne for a brief time before he was put to death by the heads of the noble families. Darayarahush (Darius I), the leader of the avengers, who sprang from a different branch of the Achaemenid family, was made king in 521 B.C. First putting down a wave of rebellion throughout the empire, he crossed the Bosphorous in 512 B.C., subdued Thrace, and

[27]

crossed the Danube, but withdrew without consolidating these gains. Aid given by the mainland states of Greece to revolting Greek colonies in Asia Minor aroused him again to action, and two campaigns against the mainland, in 492 and 490 B.C., ended in the Battle of Marathon and the Persian withdrawal to Asia Minor.

Khshayarsha (Xerxes I) succeeded his father Darius in 485 B.C. With a force of 900,000 men, supported by a huge fleet, he led a Third Campaign against Greece which culminated in the capture and burning of Athens in 480 B.C. After the defeat of the Persian fleet in the Battle of Salamis and the loss of the Battle of Platea, he withdrew to Asia Minor. Although warfare continued for some time, the Achaemenids gradually worked out more friendly relations with the Greek states and colonies.

Artakhshasa (Artaxerxes I) followed his father in 465 B.C. and reigned until 424 B.C. The first signs of the internal decay of the empire appeared in revolts in Egypt and other satrapies, and not until the accession of Artaxerxes III, who ruled from 359 until 338 B.C., were the earlier bounds of the empire reestablished for a brief period. The empire finally came to an end under the timid Darius III.

Most of our knowledge of the history of Iran in Achaemenid times and the details of the political organization, the army, and the life of the people is derived from Greek historians, especially from Herodotus. The empire was divided into provinces or satrapies, each under a satrap or governor. The governors came from noble Persian families, and the post tended to become hereditary. The basic system of an absolute monarchy and a number of semi-independent governors which was established in this period continued in Iran until the end of the nineteenth century.

The army was divided into six corps of 60,000 each, each corps composed of six divisions of 10,000 men, the cavalry mounted on horses bred in Media and armed with the bow and the javelin. The ruler's personal bodyguard was composed of 10,000 members, known as the Immortals, who were drawn from the leading families of Persia proper. The provinces were linked by roads, of which the most vital was the

"Royal Road" which ran from Susa up through Mesopotamia and Asia Minor to the city of Sardis, a distance of 1,500 miles. Messengers and travelers used a post system of relays of fresh horses stationed at the many inns along the routes. Agriculture flourished, and justice was fairly administered. Racial groups within the boundaries of the empire were allowed to retain their own religions, and often their ruling families were allowed to continue in power. The tribute in kind of early times was later developed into systematic taxation with the unit of payment the gold *daric*. Susa, Babylon, and Ecbatana were the important administrative centers and sites of royal residence, while Persepolis was the spiritual center of the empire.

The ruling aristocracy seems to have retained a chivalrous spirit, but its tribal background gradually gave way to "oriental" modes of life and manners of thought imported from the highly cultured region of Mesopotamia. Herodotus noted this tendency of the Persians toward assimilation of external influences when he wrote: "There is no nation which so readily adopts foreign customs as the Persians. As soon as they hear of any luxury they instantly make it their own." The prevailing belief of modern times that the Achaemenid Persians were barbarians in contrast to the civilized Greeks is false, for we know that Cyrus despised the commercial habits of the Greeks and it is apparent that in the fields of public administration, political organization, continuity of government, and tolerance of race and creed the Achaemenids far surpassed the Greek city states.

The documents of the Achaemenid rulers, cut into rock or baked in clay tablets, are in three languages current in the empire: Old Persian, Elamite, and Babylonian.

The royal inscriptions are limited to recording the family line and religious faith of the ruler, the names of the provinces, the suppression of revolts, and details about the construction of royal palaces. Characteristic of their pride of lineage is this passage from an inscription of Darius I in which he identifies himself as the "son of Vishtaspa (Hystaspes), Hakhamanishiya (an Achaemenid), Parsa (a Persian), Parshaya (son of a Persian), Ariya (an Aryan), cisa Ariya (of

Aryan lineage)." He also stated: "I was not a Lie-follower, I was not a doer of wrong. . . . According to righteousness I conducted myself," and, "I am not hot-tempered."

Seleucid Period

It was the fate of the Achaemenid empire to be destroyed by another world conqueror, Alexander the Great—a conquest far-reaching in its effects, for it put an end to the integrity of the ancient east and oriented it toward the west. Alexander, the son of Philip of Macedonia, was born in 356 B.C. In 336 B.C. he began to carry out his father's policies, first stabilizing the Greek mainland and then setting out toward Iran at the head of some 35,000 men. At Issus, just inland from the northeastern corner of the Syrian coast, Alexander's clever generalship put to rout an enormous Persian force commanded by Darius III, who fled when the tide of battle ran against him. Alexander then led his troops down the coast and into Egypt. Returning through Syria, he entered Mesopotamia and crossed the Euphrates and the Tigris. At Arbela he met and defeated a reformed Persian army which was ten times as large as his own. Again Darius III fled.

Alexander turned south to capture Babylon and went on to Susa and Persepolis where he seized intact the vast royal treasuries. Persepolis was burned, probably in revenge for the much earlier burning of Athens by the Achaemenid army. In the spring of 330 B.C. Alexander set out in pursuit of Darius III, first to Ecbatana and then along the southern slopes of the Alborz Range where he came upon the body of the Achaemenid king, slain by his own followers.

Scarcely pausing to consolidate his vast new territories, as though driven by a restless urge to reach the world's end, the conqueror headed for lands completely unknown to the Greeks, even into the easternmost provinces of the Achaemenid empire. The journey of discovery also had the aspect of a scientific expedition, for a careful record was kept of the line of march and distances covered, and material collected on the new peoples, plant, and animal life was sent back to Greece. In Bactria, Alexander married Roxana or Roshanak, the daughter of a Bactrian noble. Beset by nomadic warriors, he turned southward and finally crossed the Indus, and at last his soldiers, after seven long years away from their Macedonian homeland, refused to go on. Alexander agreed to turn back. The army sailed down the Indus and then made a desperate march through the barren wastes inland along the entire length of the Persian Gulf to Persepolis. At Pasargadae the violated tomb of Cyrus caught his attention, and after ordering its contents found and restored he himself sealed the door with his signet to guarantee that the tomb was forever safe from molestation. It was robbed again as soon as he left the site.

At Susa he began to disclose his plan for a new world state which would unite Macedonian and Iranian elements on a basis of equality, Hellenization of the east having already begun with the establishment of Greek colonies. Alexander himself pointed the way toward a union of the peoples when he married the eldest daughter of Darius III and encouraged 10,000 of his troops to take Persian brides. Suddenly, when he was still less than thirty-three years old, he was smitten with fever at Babylon and died in 323 B.C.

With his death, his most able generals, known as the Diadochi, or Successors, struggled for supremacy and within a few years most of his conquests were divided among the four major contestants, with each founding a dynasty. In 312 B.C. Seleucus acquired Babylon and within a few years was master of the Iranian plateau; later he took Syria and much of Anatolia. His son, Antiochus I, succeeded in 281 and consolidated the hold of the dynasty over much of the area of the former Achaemenid empire. New cities were founded and settled with Greeks, especially along the course of the ancient highway that led across the plateau to Bactria.

The Iranian plateau itself was comparatively neglected by the Seleucids whose western capital was at Antioch in Syria and its eastern one at Seleucia on the Tigris River. In Parsa a local dynasty of Persian origin emerged in the middle of the third century. These *frataraka*, or princes, considered themselves to be the successors of the Achaemenids, and used the names of Darius and Artaxerxes on their coins.

As time passed, the Seleucid dynasty became weakened

throughout by external enemies and by rivalries and open conflict among the members of the dynasty itself.

Parthian, or Arsacid, Period

About 250 B.C., a group of Iranian nomads who called themselves Parni, or Aparni, moved westwards into the former Achaemenid province of Parthava, and since they rose to prominence in a region which was known to the Greeks as Parthia, the Greeks, and later the Romans, labelled them Parthians. Shortly after becoming sedentary, they chose one of their leaders, Arshak, or Arsaces, as king, and he became the founder of the Arsacid dynasty. Reigning for an unknown number of years from 248, he led a successful revolt against the Seleucid governor. Tiridates, ?-211, established the independence of the new kingdom, and Artabanus I, 211-190, resisted the efforts of the Seleucids to reconquer it. The great leader Mithradates I, 171-138 B.C., extended Parthian rule over Bactria, Parsa, Babylonia, Susiana, and Media but allowed subject kings to retain their thrones. Crushed between the Romans and the Parthians, the Seleucid power was now broken. The Parthians then became involved in prolonged warfare against the Scythians on their eastern frontier, and migrations of the Saka about 130 B.C. caused serious difficulties which were ended by Mithradates II, 123-87 B.C., who consolidated and expanded the Parthian holdings until they stretched from within India to Armenia and took the title of "king of kings."

Under Phraates III, 70-57 B.C., began the series of wars with Rome along their common frontiers which continued intermittently for nearly three hundred years. Rome was determined to expand eastward and was interested also in the extensive commerce over the silk route whose entire western section was in Parthian hands. The fortunes of war swung east and west; Mark Antony suffered a decisive defeat at Phraata, south of Lake Rezaieh; but Trajan's later expedition into Parthia was successful.

After the triumphs of Mithradates II, the Arsacids came to consider themselves the political heirs of the Achaemenids. Their scanty cultural background was supplemented by a veneer of borrowed culture which on the upper level was both Iranian and Hellenistic. One ruler called himself a Phil-hellene, Greek modes were adopted, and for a time Greek was the official language.

As is the case for the Achaemenid period, historical information about the Arsacids comes almost entirely from outside sources, in this case Roman historians. The very name by which the rulers called their realm is unknown. They did set up a new era with year one equivalent to 248 B.C. when Arsaces became king, but there is no evidence for the existence of a centralized bureaucracy, of royal archives, or of a single command for their armies. The kings were buried, but none of the burial sites are known. The principal towns of the kingdom included Dara; Nisa, near modern Ashkabad in the USSR; Hecatompylos, the Greek name, probably near modern Damghan; and Ctesiphon, founded just opposite Seleucia in a deliberate effort to surpass the earlier site in splendor.

Sasanian Period

There is some scholarly dispute about the origins of the Sasanian dynasty. It is, however, quite generally believed that there was a local prince in Parsa, or Pars, named Sasan, who had a son, Papak, who was the head of the shrine of Anahita at Istakhr, a town near Persepolis. Papak's son was Ardashir.

In A.D. 211 Ardashir mounted a revolt against the Arsacids, and over a period of years engaged in battles which culminated with the death in battle in A.D. 224 of King Artabanus V. With the Arsacid power ended, he campaigned successfully to the north and east of his native homeland. He came into conflict with Rome, and exhausting wars with Rome and later with Byzantium continued throughout the entire Sasanian period. Shapur I, A.D. 240-272, the son of Ardashir and the second of the three foremost kings of this dynasty, at once moved west to oppose a Roman army, which, he recorded, was made up of Goths and Germans. In this fighting Gordian Caesar perished, and his successor, Philip, offered tribute and ceded the disputed territory. In the last of three later campaigns Shapur defeated an army led by Valerian Caesar and captured the Roman emperor who was taken to Susiana, modern Khuzistan, in the victor's train. Many of his army, non-Iranians who were called Aniran, were moved to the same region where, as engineers and laborers, they constructed roads, bridges, and barrages, many of which survive.

Shapur's victories were recorded on the Ka'ba-i-Zardusht in three almost identical inscriptions in Greek, Middle Persian, and Pahlavi. They begin with these words: "I, adorer of Mazda, the god Shapur, king of kings of Iranians and non-Iranians, of the race of the gods, son of the adorer of Mazda, of the god Ardashir, king of kings of the Iranians, of the race of the gods, grandson of Papak, king of the empire of Iran I am the ruler." They continue with a listing of the more than twenty lands possessed by Shapur, along with details of his campaigns against the Roman forces stating that he took Valerian "captive with my own hands."

There were nearly thirty rulers of this line. Narse, A.D. 293-302, was less successful against Rome, and the Tigris River became a fairly stable boundary between east and west. Shapur II, the great-great-great grandson of Ardashir, came to the throne in A.D. 310 and during his reign of sixty-nine years waged three separate wars with Rome. Hearing that they were ready to sue for peace, he sent a letter to the Emperor Constantius II which was recorded by the Roman historian Ammianus Marcellinus. It opened with these words: "I, Shapur, King of Kings, partner with the Stars, brother of the Sun and Moon, to my brother Constantius Caesar offer most ample greeting." It continues, ". . . it is my duty to recover Armenia with Mesopotamis, which double-dealing wrested from my grandfather. That principle shall never be brought to acceptance among us, which you so exultantly maintain, that without any distinction between virtue and deceit, all successful events of war should be approved." In conclusion he wrote: "I surpass the kings of old in magnificence and array of conspicuous virtues. But at all times right reason is dear to me, and trained in it from my earliest youth, I have never allowed myself to do anything for which I had cause to repent."

The earliest of the Sasanian kings chose as their successors, usually, but not always, one of their sons. By the time of the reign of Shapur II the king was elected by the highest dignitaries of the realm—by the leading religious figure, the *mobadan mobad*, by the chief military leader, *Iran-spahbad*, and the head administrator, *Iran-diberbad*. Prior to his demise, the king wrote letters in his own hand to these personages giving his observations on the character and abilities of the candidates. In due course these letters were opened, with the *mobadan mobad* presiding over the election and then crowning the new king.

Khusraw I, popularly known to later ages as Anushirvan the Just, came to the throne in 531, and reigned throughout the most brilliant period of the Sasanian dynasty. It was marked by his stern yet benevolent rule for some fifty years. At home he initiated social, fiscal, and military reforms.

By the time Khusraw II succeeded in 589, the threat to the Sasanians was from Byzantium rather than Rome. Although in 628 this reign ended in disaster, with the ruler slain by his own commanders, it was during his reign that the splendor of the monarchy was at its height as regards regal pomp and the display of fabulous treasures.

As early as A.D. 611 Arab forces had met and defeated Sasanian troops. After the Arab conquest of Syria, the Muslim armies turned towards Iran: in A.D. 637 a Sasanian army was annihilated and Ctesiphon taken. In A.D. 642 at Nihavand the Arabs gained a final crushing victory. Yazdagird III fled to the northeast where he was murdered by his own followers— shades of Darius III—and the dynasty ended.

The Sasanian period witnessed the rebirth of a nationalistic Iran, strong and prosperous in its own right, and unreceptive to foreign contacts and influences. The divine right of the kings was reflected in scenes of the investiture of the monarchs by the gods. The rulers were proud to serve and to adore the gods, although they too were gods. The basic distinction between Iranians and the Aniran, or non-Iranians, or, as in some inscriptions, between Aryans and non-Aryans, was maintained. Outsiders were ill regarded, so that it is quite curious that the victorious Arabs called the Iranians *ajami*, foreigners or barbarians, just as the Greeks called the Iranians of Achaemenid times barbarians. As in Achaemenid times, most of the detailed material about the realm comes from outside sources, in this case early Arab historians.

The remarkable stability of the empire over centuries was largely the product of an efficient and highly centralized administration, with its ministers and numerous secretaries. The army was paid and controlled by the royal treasury.

The top rungs of the social ladder were held by the king, then the *shahrdar*, or subordinate kings in the lands of the empire, then great nobles, and, finally, lesser nobles. Many of their names are known, and in translation are quite close to the honorary titles found in the later history of Iran.

It is said that there were well defined social classes, but the material on these groups, some of it from Arab historians, is by no means precise and not without some ambiguity. Supposedly, society was divided into four classes: priests, warriors, secretaries, and commoners. The first three groups were headed by officials of several grades, while the commoners included farmers, merchants, and artisans. Basically, society was sharply divided between the few highly privileged and the many underprivileged.

The center of the empire was the city of Bishapur, in the land of Fars, but two other major cities, Ctesiphon and Gundeshapur, were in other areas of the empire.

Religious and secular festivals marked the passing year. New Year's day, called No Ruz, lasted for six days. The ruler received the nobles of the realm who brought him presents, and they received rewards in return: gifts of money, robes of honor, and appointments to lucrative posts. Token taxes were offered to the ruler, money was newly minted, and the fire temples were purified. Poets celebrated the occasion in verse, and the public at large was admitted to a royal audience.

The Arab Conquest of Iran

Early in the seventh century Khusraw II captured Damascus and Jerusalem and advanced to the very gates of Constantinople. Then the tide turned and the Byzantine Emperor Heraclius moved across Mesopotamia as far as Ctesiphon. At last Iran and Byzantium signed a peace of sheer exhaustion.

Just then came the Arabs, overrunning Iran and bringing with them the Muslim religion. The centuries from the Arab conquest to the present day are generally known as the Islamic period of Iran; they cover almost exactly the same number of years as lay between the rise of Cyrus and the end of the Sasanian empire.

Arabia has always been the home of nomads, a Semitic people hostile to foreign invasion and influences, and in the opening years of the seventh century the peninsula was quiet. Medina and Mecca were small towns owing their limited prosperity to their positions along a main caravan route, and Mecca had been for some time the center of pagan cult worship and a place of pilgrimage. Upon the quiet scene appeared the Prophet Muhammad, born in 570. His great mission was not apparent until he had reached middle age, when the Qoran was revealed to him, and his religion gained momentum slowly. He made his exodus from Mecca to Medina in A.D. 622, the date which later became the Muslim year one, and attained complete control over the peninsula of Arabia only a few years before his death in A.D. 632.

His devoted friend and early follower, Abu Bakr, became the first *khalifa* (Caliph), or "Successor," and led a series of plundering raids into Mesopotamia and Syria, with such success that the nomadic warriors soon came into contact with the major forces of Byzantium and of the Sasanians. Heraclius was defeated in Palestine in A.D. 634, and in 637 a Sasanian army 120,000 strong was defeated in the four-day-long battle of Qadisiyya. Ctesiphon fell, and the Arabs spread out onto the Iranian plateau. The Arabs overturned governments which were in a state of corruption and decay and the hope they held out to the great masses of the people of more equality and kinder treatment brought on a social and religious revolution. It is true, however, that there was no serious attempt at wholesale conversion to the Muslim religion. Much of the population of Iran converted to Islam in self-interest. On the one hand, to escape the poll tax levied on the *zimmis*, or infidels, and, on the other, as the necessary first step in advancement under the new authority.

Within Iran the conquest spread, spurred on by the Caliph

'Uthman's promises of the governorship of wealthy Khorasan
to the first of his generals to reach the province. In A.D. 652
Iranian forces were defeated at Khwarazm on the Oxus River,
and within a few years lands in distant Asia were under the
Muslims. Bands of Arabs moved to Khorasan and settled
there. The famous general Qutayba, who campaigned in the
area from 704 until 715, was so zealous in promoting the
Muslim faith that he quartered Arabs in every household of
the captured towns and at Bokhara built a large mosque on
the site of a fire temple. The Arabs, as they advanced across
the plateau, entered into separate treaties with towns and with
feudal lords—many of these lords were the Sasanian *dihqan*
who held small fortresses. Garrisons of Arabs were in the
towns, and taxes were collected for their upkeep, but social
life and commerce were probably but little altered.

For more than a century after the rise of Islam all the conquered lands gave unquestioning obedience to the authority
of the Umayyad Caliphs, resident in Medina or Damascus.
The areas were divided into provinces with a military governor and a head-tax collector in the highest station in each
province. Details of administration, especially the complicated
task of collecting taxes, were left in the hands of the officials
of the regimes displaced by the Arabs. In Mesopotamia two
tax registers were kept; one in Arabic for the new overlords,
and one in Persian for the native population. Difficulties of
communication meant that the more distant provinces were
less firmly controlled by the central authority, so that Iran
was not too severely treated. However, the coming of Islam
did produce profound alterations in the political, economic,
and social structure. The age-old Iranian ideas of divine right
and autocratic control were challenged by a democratic spirit
and by the internationalism of the new religion, while at the
same time the cultural superiority of the Iranians and their
pride in their institutions remained, to stamp the cultural and
artistic future of Islamic Iran with a character quite different
from that of any of the other Muslim countries.

In the eighth century strong feeling arose against the
Umayyad Caliphs, and in A.D. 750 the reigning Caliph was
defeated in battle. A new line of Caliphs known as the 'Abba-

sids took over the power, and in 763 founded a famous circular city at Baghdad which was to be their capital until 1258, when the Mongols sacked the city and ended the 'Abbasid line. Iranian soldiers had aided in the overthrow of the Umayyads, and in the new court Persians held high positions and Persian dress and manners prevailed.

The Rise of Local Dynasties in Iran

Quite soon after the Arab occupation of Iran, local dynasties emerged in two areas. One was the Caspian Sea littoral and its hinterland where the broken topography made movement difficult and where invading forces could be avoided by retreating into the wooded mountains. The western end of the Caspian Sea region was called Daylam, its eastern end was Tabaristan. The other area was that of the province of Khorasan, "land of the sun," in northeastern Iran.

The Baduspanid dynasty of 53 rulers held a chain of fortresses in Tabaristan from about 45/665 until—as incredible as it may seem—1006/1597.

The Bawandid dynasty was founded by Baw, a Sasanian military governor of Khusraw II. Also centered in Tabaristan, it endured from 45/665 to 750/1349, because of the willingness of its lords to become vassals of more powerful lines.

The Dabuyids, a line founded by Dabya, were in close relations with the Bawandids from 50/670 until 116/734, and remained unconverted to Islam. One of them, Khurshid, called Farshwad Marzuban, or "warden of the frontier," reigned in splendor. Historians record that he was a great builder and a keen hunter. His private life may have been complicated by his 93 wives, each with her own palace: one of them was Giran Gushwar, "she of the heavy earrings." His tribute to the Caliph at Baghdad included gold and silver, green silk carpets, colored cottons, gold-embroidered garments, and saffron. These rulers considered themselves heirs of the Sasanians, with their coins repeating those of Khusraw II.

From the point of view of political and cultural influences, the Ziyarid dynasty was the most important of the kingdoms and principalities of the Caspian Sea littoral. It was in power from 315/927 until about 483/1090 in Tabaristan and further afield. Its founder Mardawij, "who attacks people," ibn Ziyar carved out his kingdom by force of arms, and in the seven years before his assassination in 323/935 campaigned successfully as far south as Isfahan. There he took his seat on a throne of silver covered with silver and gems, put on his head a crown topped with pearls and sapphires, and said: "I shall restore the empire of the Persians (al-'ajam), and destroy the empire of the Arabs." He stated that he was another King Solomon, with his Turkish slaves as his demons, and ordered that he should be addressed as the king of kings.

Mardawij was of a rather difficult temperament. As a non-Muslim, he ordered a magnificent celebration of the Sasanian festival of fire, called Sadagh. Fireworks were prepared, and some 2,000 crows, bearing lighted torches were to be released. Dissatisfied with the scale of the preparations, he insulted his retinue, with the result that he was murdered in his bath that same night.

Two later rulers of this line were outstanding personalities. Shams al-Ma'ali ibn Vushmagir, "snarer of quails," Qabus reigned for 35 years starting in 367/978. He studied philosophy and astronomy, composed poetry in Persian and Arabic, and was an excellent calligrapher. To his court he welcomed such renowned figures as al-Biruni and Ibn Sina (Avicenna). His well preserved tomb will be described in a later section.

'Unsur al-Ma'ali Kay Ka'us ibn Iskandar came to the throne in 441/1049 and crowned his career with the writing of the *Qabus nama*, in which he traced the family back to a Sasanian governor of the time of Khusraw I, and offered advice to his son.

The Buyid dynasty had its origins in Daylam. Members of the family accompanied Mardawij ibn Ziyar on his campaigns, and at the time of his assassination the eldest of three brothers, 'Ali ibn Buya, commanded his garrison at Isfahan. He moved south to take Fars and founded the dynasty in 320/932. His brothers acquired central Iran, Kerman, and Khuzistan. A few years later his brother Ahmad marched to Baghdad. The caliph met him at the city gate and conferred a title on him. A few days later Ahmad was received in audience and kissed the earth and then the hand of the caliph. Suddenly his followers burst into the hall, pulled the caliph from his throne, and dragged him to Ahmad's quarters. A new caliph was named and the Buyids dominated Baghdad for some years.

'Adud al-Dawla Fana-Khusraw, who came to the throne when he was about fifteen years old, reigned from 338/949 until 372/983, and held on to most of the territories acquired earlier. In addition, his forces crossed the Persian Gulf to conquer Oman. Best known of this line, his activity centered in Fars, and at Persepolis are two inscriptions carved for him in Arabic telling of his victory at Isfahan and the fact that a person was presented to him who translated the ancient inscriptions on the monuments.

When Majd al-Dawla inherited from his father at the age of nine the administration was taken over by his energetic mother. When Mahmud of Ghazna ordered her to mint coins in his name or suffer the consequences, she replied: "The fate of armies is uncertain. If the sultan beats me, the victory over a widow will not bring him great glory; if, on the contrary, he should be defeated by me, the stigma of this disgrace will not disappear from the brow of his fortune until the last judgment." After his mother's death Mahmud's soldiers carried off Majd al-Dawla into captivity and the main line came to an end.

The Kakuyid dynasty stemmed from Rustam ibn Marzban Dushmanziyar, who was granted a small fief by the Buyids. His last name gives the clue to his reward: it means "enemy of the Ziyar," commemorating the support he gave the Buyids against the Ziyarids.

His son, 'Ala' al-Dawla Muhammad, was also known as Ibn Kakuya. In childish speech *kaku* is used for maternal uncle, rather than the correct words, *khalu* and *daii*, and he seems, in fact, to have been the maternal uncle of the Buyid Majd al-Dawla. In 398/1008 the Buyids placed him in charge of Isfahan; he became virtually independent in western Iran. His successor, Abu Mansur Faramuz, was forced to become a vassal of the Seljuqs, but he and his successors managed to

[41]

cling on to the town of Yazd for over a half century longer, that is, until 513/1119.

In this review of the dynasties of the Caspian Sea littoral, it was apparent that some of their rulers moved onto the plateau proper and in these accounts the text has fallen, so to say, behind in chronology. Turning to the region of Khorasan, more important dynasties appear and chronology is set right.

The Tahirid dynasty was founded by Tahir ibn al-Husayn, called Dhu'l-Yaminiyn, the "Ambidextrous," who was sponsored by the Arab governor of Khorasan. As a military commander, he found favor with the caliph and was stationed in Baghdad. Later he was ordered back to Khorasan, where he and his successors, Talha ibn Tahir and Abu'l 'Abbas 'Abdullah, were semi-independent rulers with their capital at Nishapur. 'Abdullah had a special interest in the welfare of the farmers and ordered the writing of the *Kitab al-Quniy*, or Book of Canals, which codified the regulations covering the water distribution for irrigation. He also encouraged education for all classes, both the "worthy" and the "unworthy."

Muhammad ibn Tahir lost Nishapur to the Saffarids, and the dynasty which lasted from 205/821 until 248/873 came to its end.

The Saffarid dynasty sprang from Ya'qub ibn Layth and his three brothers, one of whom, 'Amr, succeeded to his power. The brothers had come to Zaranj in Sistan, with Ya'qub hiring out as a coppersmith, *saffar*, and 'Amr as a muleteer. The brothers formed a robber band, which gained a reputation for generosity to people of their own kind. By 871 Ya'qub had taken over Khorasan from the Taharids and campaigned successfully as far as Herat and Kabul in the north and east, and Fars in the west. After he defeated Muhammad ibn Tahir the caliph felt compelled to recognize him as viceroy of Khorasan, Tabaristan, Ray, and Fars. With larger ambitions, he marched on Baghdad, but was defeated and died in 266/879.

'Amr professed loyalty to the caliph and was confirmed in possession of the regions named. From his seat at Nishapur large amounts of tribute were forwarded at regular intervals to Baghdad. His effort to expand his hold to include Transoxiana led to his downfall: defeated in battle by Isma'il the Samanid, he was sent as a prisoner to Baghdad. The Saffarids were down but not out, for the family was to survive in Sistan for nearly six centuries longer. As vassals of the Ghaznavids, Seljuqs, and Mongols, they managed to retain the personal loyalty of their followers.

The Samanid dynasty is said to have descended from Saman-khuda, headman of the village of Saman near Balkh, who himself claimed descent from Bahram Chubin, a Sasanian commander. His grandson, Ahmad ibn 'Asad ibn Saman, actually initiated the dynasty, after he and three of his brothers had served as sub-governors within Khorasan. Ahmad established firm control over Transoxiana, and passed on his realm to his son Nasr who ruled from Samarqand under a diploma of authority from the caliph. Isma'il ibn Ahmad, the best known figure of this line, seized power from his brother. As mentioned earlier, he defeated 'Amr, the Saffarid, and then went on to take over much of the Caspian Sea littoral and other northern areas of Iran. Under Isma'il the Samanids were the strongest power in eastern Iran, and their authority reached as far as India.

In this period we are on firm ground as concerns knowledge of the organization of government, and the details of its bureaucracy. The *vazir* was the chief official of the realm: his insignia of office was an inkwell. Under his direction the affairs of state were conducted by some ten *divan*, or offices. To mention only one of these, a postmaster supervised the mails which were for official use only.

By the middle of the tenth century the Samanid state had begun to disintegrate. Internal palace revolutions and rebellions in Khorasan ushered in the final collapse of the dynasty at the end of the century.

The Ghaznavid dynasty was a Turkish line whose fortunes rose as those of the Ziyarid, Samand, and Buyid declined. Under the Samanids a series of Turkish slave governors, or *amirs*, were entrusted with Khorasan, Ghazni, and other areas. In 961 Alptigin, amir of Khorasan, broke away from the Samanids, besieged Ghazni and took it for himself. After a few years of confusion Abu Mansur Sebuktigin acquired Ghazni, and for nearly twenty years remained loyal to the Samanids, becoming independent only near the end of his life. Sebuktigin, the first of the Ghaznavid line, was the author of a brief *Pand nama,* or "Book of Counsel," written for his son Mahmud who was then, at age seven, his deputy at Ghazni. Recalling his own life, he wrote that when he was twelve he was seized in a raid by a neighboring Turkish tribe and sold to a slave dealer. After training as a soldier, he was bought by Alptigin under whom his fortunes prospered. Sebuktigin initiated the prosperous raids into the plains of India which Mahmud was to turn into full-scale campaigns. From seventeen such campaigns between 1000 and 1029 Mahmud brought vast treasures back to Ghazni. Closer to home, he established the Oxus, now the Amu Darya, River as the northern frontier of his realm, took Khorasan, and penetrated into western Iran to seize it from the Buyids. He was a military genius, but he did not neglect literature and the arts and spent vast sums to embellish Ghazni, his capital. His empire, assembled in such a few years, began to fall quickly apart after his death in 1030. Khorasan and other areas were lost to the Seljuqs, and in 1150 a conqueror set Ghazni ablaze. Finally, the dynasty was brought to an end in 1186.

Seljuq Period

The Seljuqs first appear in the history of Iran as nomadic warriors who crossed the Amu Darya River in the early eleventh century and moved westwards. The Seljuq family was the ruling element of the Qiniq group of the Oghuz, or Ghuzz, Turks. After prolonged contacts with the Samanids and the Ghaznavids, about 1045 they moved into Khorasan and stood off attempts to drive them out by larger, better armed Ghaznavid forces. In 1039 Merv, Herat, and Nishapur surrendered to them, and in the following year they defeated a great host led by Sultan Mas'ud of Ghazni.

Toghril, the first ruler of the line, established himself first at Nishapur and later at Ray, and assumed the title of Exalted Sultan. Initiating diplomatic correspondence with the caliph, he marched on Baghdad. Arriving there in 1055, he was kept waiting for a long time by the caliph, but the waiting was worthwhile for he was given seven robes of honor and two crowns to go with his new title of King of the East and of the West. Now the secular power of the caliphate was taken over, in large part, by the strongest kingdom of the region—a situation which was to prevail in all later periods.

The main line of the Seljuqs was to rule for one hundred years from the Bosphorus to Chinese Turkestan; minor branches held on for a still longer period.

At the time of the advent of the Seljuqs into central and western Iran much of it was held by petty rulers ready to fly at one another's throats. The Seljuqs overthrew the local kingdoms, united Muslim Asia, and embraced Islam with great fervor. This period became one of the most momentous in the long history of Persian culture, for it was then that the rough, illiterate Turks first placed Persians in the highest official posts and then themselves developed into patrons of learning and the arts.

In 1063 Toghril died without male issue, and was succeeded by a nephew, Alp Arslan. His reign of ten years and that of his son Malik Shah for twenty more were times of great commercial activity. In 1071 Alp Arslan met the Byzantine forces in Armenia, defeated them, and captured the emperor Romanus Diogenes who was permitted to ransom himself from captivity. The following year Alp Arslan was stabbed by a local lord whom he had condemned to death. Just before he expired he remarked: "Yesterday, when I stood on a hill and the earth shook beneath me from the greatness of my army and the host of my soldiers, I said to myself, 'I am king of the world and none can prevail against me; wherefore God Almighty has brought me low by one of the weakest of his creatures. I ask pardon of Him and repent of this my thought.'"

Under Malik Shah the limits of the kingdom were extended—into Syria, Palestine and Anatolia, and in Arabia to Bahrain and as far as Yemen. He established the astronomical observatory in which 'Umar Khayyam and other men of science were employed to make the calculations for a new calendar, and adorned Isfahan, his capital, with many fine buildings and gardens. His personality, and even his achievements, were overshadowed by the activity of his vazir, Nizam al-Mulk, who was assassinated just two months before the ruler's own death in 1092.

As did so many notable Iranians, Nizam al-Mulk, the "Regulator of the Kingdom," came from Tus in Khorasan. Born Abu 'Ali Hasan ibn 'Ali Tusi in 1019, he studied theology at Nishapur, and at age twenty-two took employment as a secretary at the court of Ghazni. Later he entered the service of the Seljuqs and rose to the post of vazir under Alp Arslan and Malik Shah. He headed the administration, the *Divan-i-vazir*, which effectively ran the state, the ruler and his court playing no active role in daily affairs. One of his major interests was education, and he established a *madrasa*, or religious school, in each of the major towns: in his honor they were called Nizamiyya. At the request of Malik Shah for a report on the causes of the weaknesses of the Seljuq kingdom he wrote the *Siyasat nama*, "Treatise on Government"; this work is described in a later section.

Malik Shah's death was followed by internal struggles for the throne which ended only after the death of his son and immediate successor, Berk-Yaruq, in 1105. Next in line was a half-brother Muhammad who was the last of the dynasty to maintain firm control over Iran and Iraq.

In 1118 Mahmud succeeded his father Muhammad and ruled until 1131. His brothers held power in several parts of the kingdom, while his uncle, Sanjar, long the uncrowned ruler of Khorasan, pushed westwards. Mahmud was followed briefly by his brother Toghril and then by another brother, Mas'ud, whose reign, ending in 1152, was marked by almost continual intrigue, warfare, and rebellion. Sanjar then came to the throne after a half century of effort.

As on many earlier occasions, tribes from remote Central Asia were trying to push across the Amu Darya River, and Sanjar was hard put to hold back the Qara-Khitai tribe. Moreover, in 1153 Sanjar was captured at Merv by nomadic tribesmen, and held prisoner for three years in degrading conditions. A year later, in 1157, Sanjar died and was buried at Merv in a great mausoleum built during his lifetime. With his death the authority of the Seljuqs in eastern Iran collapsed, although the line did not become extinct in Iran until 1194 with the death of Toghril in battle—the line of the so-called Great Seljuqs began and ended with rulers named Toghril.

At first the Seljuq armies were composed largely of slaves, but then they took on a professional character. Recruited bodies of Turks, Armenians, Greeks, Kurds, and Daylamites formed their core, and could be augmented by tribal auxiliaries. The professional soldiers were rewarded with the income from grants of farm land. Taxation lay heavily on the land, and as the central authority weakened tax collections fell off so that a ruler's treasury might stand empty.

The Seljuqs were plagued by the Muslim sect of the Isma'ili who maintained a virtual state of their own in Iran from 1090 until 1256. The tenets of this sect will be discussed later, here the concern is with their temporal activity. Each area they held was headed by a *da'i*, or head of religious instruction. In 1090 they set up their headquarters under their chief *da'i*, Hasan-i-Sabbah, at the isolated fortress of Alamut to the north of Qazvin. There he held sway until his death in 1124. Other key fortresses throughout Iran were seized from their local garrisons. For these fortresses and from secret cells in the towns, *fida'i*, or devotees, were dispatched to murder important persons who, as military commanders or figures in administration, represented a threat to the Isma'ilis. Often the attacks were made in public, so as to spread fear of the sect, and it claimed responsibility for the killing of Nizam al-Mulk. Also active in Syria, they were there known as the *Hashishiyya*, or hashish smokers, and this name the Crusaders turned into Assassin. The story that the devotees had visions of Paradise under the influence of hashish, were then told that they would go directly to Paradise if they killed on their missions, and were then sent on these missions high on the drug, lacks any substance.

Mongol Period

Along the lower course of the Amu Darya River lay the irrigated agricultural region of Khwarazm. Under the Seljuqs the family of a Turkish slave became its hereditary governors, and acquired the title of Khwarazm-Shahs. Following the downfall of the Seljuqs, in the opening years of the thirteenth century, 'Ala' al-Din Muhammad Khwarazm-Shah found himself master of a kingdom which stretched from India to Anatolia.

About the year 1160 occurred the birth of Temujin, who under the title of Chingiz Khan was destined to lead the Mongol hordes across the breadth of Asia. Rising from utter obscurity, Chingiz marshaled together the pastoral nomadic clans and swept south and east to overrun civilized China. In 1219 he turned west at the head of some 700,000 men, many of them mounted on wiry steeds. At the limits of the Khwarazmian kingdom he paused to send envoys to Sultan Muhammad. The latter scorned to treat with these unknown savages, and the result was the speedy destruction of his realm.

The Mongol army moved ahead to take Bokhara, Samarqand, Balkh, and Merv. Towns which offered resistance were besieged, stormed, burned, and frequently obliterated from the face of the land. Nishapur fell in 1221, its inhabitants and all living things, including cats and dogs, were slaughtered. Occasionally artisans were spared. Indeed, across the troubled centuries one's best hope of survival lay in being a coppersmith, mason, carpenter, painter, or other craftsman. The slaughter was so extensive and the destruction of irrigation works and towns so widespread that many areas did not recover for centuries.

Jalal al-Din, successor of 'Ala' al-Din, led a valiant resistance, and was hunted across the country to the shores of the Caspian Sea where he escaped in a small boat just ahead of his pursuers. Chingiz Khan returned to the East when almost all of Iran had been overrun, and died there in 1227. The council which named his successor determined to send an army against the remnants of the Khwarazmian power, and the general Charmaghan led the Mongols as far as northwestern Iran and Iraq.

In 1245 Hulagu, a grandson of Chingiz, led a new wave, with orders to take and destroy the fortresses of the Isma'ilis. Historians tell of the negotiations between the Mongol leader and the head of the sect which resulted in the surrender of some of the fortresses. Others, including Alamut, were captured and razed. Hulagu then marched to Baghdad: in 1258 it was taken by storm and the Caliph Musta'sim and his family wiped out. Turning towards the lands of the Egyptian Mamluks, Hulagu's forces took Damascus, but, in a battle vital for the future of this part of the world, were defeated by the Egyptians at 'Ayn Jalut in Palestine. This was the first check suffered by the Mongol forces.

The remaining years of his life Hulagu spent in and around Azerbaijan, moving nomad-like with the seasons, and possessed of the title Il Khan, or subordinate khan, awarded him by the great khan, Kubilai. This title was passed on to his successors so that his dynasty is known as the Il Khanid. Hulagu was buried on the island of Shahi in the salt lake now called Lake Rezaieh; his funeral seems to have been the last time in Iran when the attendants of a Mongol dignitary were immolated with their master.

Once the Mongols had settled in Iran they were subject to manners, modes of dress, a religious belief, and a culture foreign to their tribal way of life. The force and continuity of Iranian civilization worked to alter their very character. The task of putting together shattered Iran was undertaken with the cooperation of the experienced local bureaucracy, the efficient traders, and the Muslim clergy: it included the imposition of stern control over feudal lords and nomad chiefs.

Abaqa succeeded Hulagu in 1265 at the age of thirty-five. The military strength of the realm may have passed its peak with the withdrawal of large forces to Mongolia, and his armies were not too successful against the Mamluks. After one defeat Abaqa sent this word to the Sultan Bibars: "You pounced like a robber on the advance guard of our army and have defeated it, and when we drew near you fled like a thief." The Crusaders were under heavy attacks from the Mamluks and agreed with the Mongols on the urgent need for a union against Egypt: on one single occasion they actually joined forces in an attack on Caesarea. Joint action was the subject of correspondence between Abaqa and Pope Clement IV and Edward I of England, himself a Crusader.

Abaqa met his death in 1282. At a feast he became very

drunk, wandered into the garden and shouted to his guards to shoot their arrows at a big black bird that was sitting on a branch and menacing him. They saw no bird, while Abaqa fell into a faint from which he never awoke. The historians of the time praised him, only deploring that he remained an idol worshipper, and noting that he appointed a just man to protect the weak and the poor. Believing that theft was the result of poverty, he often spared those who had stolen from him.

According to his will, Abaqa wished to be succeeded by a son, Arghun, but the Mongol code took precedence and a brother, Takudar, was chosen. He immediately professed Islam, and took the name of Ahmad. Fortunate chance has preserved a long letter which he wrote to Sultan Kelavun of Egypt, along with the latter's even longer answer. Ahmad wrote that his amirs and generals wanted to attack Egypt with "such a multitude of men that the earth could scarcely hold them," but that he disagreed because there should not be discord and enmity among Muslims. His amirs and generals were incensed by his action, and plots against the ruler began to multiply. His short reign ended in 1284, when, on orders from Arghun, he was seized and killed by having his back broken.

Arghun soon declared war on Egypt and resumed the exchange of envoys and letters with Europe: these letters were in a dialect of the Mongol language peculiar to Iran and were written in Ouigur characters. With the Mongol court established at Tabriz, that town began to grow and prosper: its foreign colony included merchants from Pisa and Venice. The administration of the kingdom was in the hands of very capable, cultured Persian families. Unfortunately, these vazirs were the objects of plots, and were never able to retain the trust of the rulers. One after another they were put to death, and it was not until 1324 that one of them died from natural causes.

Arghun was a man of great strength, able to vault over two horses to land in the saddle of a third. He was a firm believer in alchemy, and supported many practitioners of that art. He stated that he knew that some of them took advantage of him, but that if he were to mistreat any such the one person who knew the secret would be afraid to come to his court. One of their number offered him a potion guaranteed to secure a very long life. A compound of sulphur and mercury, it may well have been the cause of the lingering illness which, in 1291, ended in death. Earlier Arghun had sent a messenger to the Great Khan, asking for a new bride from the family of his principal wife who had died. A charming maiden of seventeen was selected and entrusted to the Polos, brothers and son, to escort to Tabriz. Before they reached Tabriz, described by Marco Polo as a "great and noble city," Arghun was dead, and the girl was handed over to his son Ghazan.

Gaykhatu, a brother of Arghun succeeded, and ruled for four years. His reign was marked by an attempt to introduce paper currency modeled after that in use in China: the bazaars closed down, trade ceased, and chaos resulted. In 1295 he was strangled with a bowstring and succeeded for a brief reign of a few months by a relative, Baidu.

The accession to the throne in 1295 of Ghazan, a great-grandson of Hulagu, marked the opening of a period of progress and prosperity. Ghazan, a Muslim, took the name Mahmud, and from this time until the collapse of the dynasty Islam was the religion of the state and Iranian culture played a dominant role. Through his Persian vazirs he instituted a series of measures designed to bring about general prosperity. The activity of the towns was favored by lower taxes on the trades and crafts. Intensive farming was promoted through measures to extend irrigation and promote the cultivation of neglected land, while the farmers were protected from violence and the forced quartering of soldiers and officials on their lands. Ten thousand men guarded the main roads along which numerous caravans passed in safety.

Ghazan Khan was said to be short and unprepossessing in appearance, even ugly. One of his first decrees called for the destruction of all churches, synagogues, and Buddhist temples, but within a few years he showed more tolerance to these faiths. Testifying that he was a sincere Muslim, he assembled his amirs and they exchanged their broad-brimmed Mongol hats for Muslim turbans. In 1299 his armies inflicted a smashing defeat on the Egyptians at Homs in Syria, and moved on to take Damascus. Failing to consolidate this victory, they were later driven out of Damascus by the Egyptians.

At Tabriz his building activity centered in the suburb of Shenb, begun by Arghun, and featured his own mausoleum which was well on the way to completion at the time of his death in 1304.

For most of his reign Ghazan enjoyed the fruitful services of a remarkable vazir, Rashid al-Din Fazl Allah, a Persian born at Hamadan whose earlier career had been that of a physician. An eager patron of the sciences and the arts, he was also a prolific writer, his outstanding work being the *Jami' al-Tawarikh*, a universal history which he completed in 1310.

Oljeitu, a brother of Ghazan, came to the throne. Baptized as a Christian, he later embraced Islam and took the name Muhammad Khudabanda. As a Muslim, he was at first a Sunni and later a Shi'a. His major undertaking was the construction of a new capital city, Sultaniya, in a fertile plain not far from Qazvin. Between 1307 and 1313 a citadel, mosques, palaces, and residential quarters were completed, while the focal point of the site was his own mausoleum. Dying in 1316, he was placed in this structure in a sarcophagus covered with jewels. General mourning was held for eight days, with the people reviving an ancient Iranian custom by donning blue garments.

His son, Abu Sa'id, mounted the throne at Sultaniya at age twelve in the presence of nobles and dignitaries who knelt on the ground with their heads bare. Earlier his father had named him governor of Khorasan: accepting the appointment Abu Sa'id declared: "I pray your Majesty to believe that notwithstanding my young age [he was then nine], I possess the proper spirit and tact needed by a leader of men." On the edges of the kingdom amirs sought to take advantage of his youth, and insurrections broke out. Order was restored with the ruler leading his forces in battle and acquiring the title of Bahador, the Brave.

With the death of Abu Sa'id in 1335 a large number of pretenders to the throne appeared. Few had strong claims, since the direct line of Hulagu had thinned out over the years. Two of them ruled briefly before the kingdom broke up. Several regions now enjoyed independence. Some were held by dynasties established in the Seljuq period or under the Il Khanids, while two of these dynasties survived until the Timurid period.

Salghurid Dynasty

This so-called Atabeg, or father-lord, line, came from a Turkman tribe called Salghur or Salur, which took service under Toghril, the Seljuq ruler. The founder of the dynasty, Sonqur, managed to acquire Fars, with Shiraz as his capital. His son, Zangi, was confirmed in possession of the area by Toghril.

The poet Sa'di took his *nom de plume* from the next of the line, 'Izz al-Din Sa'd, who became a tributary to the Khwarazm-Shahs. His son Abu Bakr quickly yielded to the Mongol invaders and was awarded the title Qutlugh Khan. Indeed, the conciliatory attitude of these rulers towards invaders succeeded in sparing Fars from the devastation which swept across more northern Iran. The successors of Abu Bakr each reigned very briefly until a granddaughter of his, Abish Khatun, came to power. This power she soon yielded to her new husband, a son of Hulagu.

Muzaffarid Dynasty

Sharaf al-Din al-Muzaffar, a commander of the Mongols, was named by Oljeitu as governor of the region between Isfahan and Yazd. In 1314 his son, Mubariz al-Din Muhammad, succeeded to that post, and a few years later Abu Sa'id added that of governor of Yazd. He added Kerman to his holdings, and was successful in defeating the Inju' ruler. According to contemporary historians, Muhammad was cruel and bloodthirsty. One of them wrote: "I often saw culprits brought before Muhammad when he was reading the Qoran. He used to stop reading, kill them with his own hand, and then resume his pious occupation." His harsh treatment of his sons led two of them to have him seized and blinded.

Jalal al-Din Shah Shuja', a son of Muhammad, managed

to gain ascendancy after months of conflict. In fact, his entire reign was strewn with battles, and with as many defeats as victories. He sent rich presents to Timur, then advancing through Khorasan, as well as a letter in which he commended his sons and family to the conqueror.

Shuja' died in 1382, and his son and successor, Mujahid al-Din Zayn al- 'Abidin 'Ali, having the poor taste to arrest a courier from Timur, soon fled before the approach of Timur to Shiraz. A few years later Timur returned to Shiraz and in 1393 ended the Muzaffarid line by executing all the remaining princes.

Inju' Dynasty

In 1303 Sharaf al-Din Mahmud Shah Inju', with his last name meaning "tax collector," was named administrator of the Il Khanid property in Fars. By 1325 he had become a virtually independent ruler at Shiraz, but he was assassinated in 1335. Jalal al-Din Mas'ud who succeeded was driven from power by a brother, who was killed by another brother. In his turn, this brother was assassinated, and Jalal al-Din Mas'ud came to the throne a second time until he was also assassinated in 1343. So many plots, feuds, and murders were crowded into a short space of years that little else remained to be recorded.

The successor of Mas'ud, Jamal al-Din Abu Is'haq, is the best known of this line, in part because of his longer reign and in part because of the favorable artistic climate of Shiraz in these years. Two writers, 'Ubaid-i-Zakani and Hafiz, whose fame was to wax in later centuries were active there.

Abu Is'haq's end was as violent as those of his predecessors for he was captured and executed by Mubariz al-Din Muhammad in 1357. Too addicted to poetry and pleasure, when told of an enemy advance on Shiraz he said, in verse, "Come, for just one night we take pleasure together, and on the morrow deal with tomorrow's business."

Jalayirid Dynasty

The Jalayir was a Mongol tribe, some of whose members came to Iran with Hulagu, and were themselves directly re-

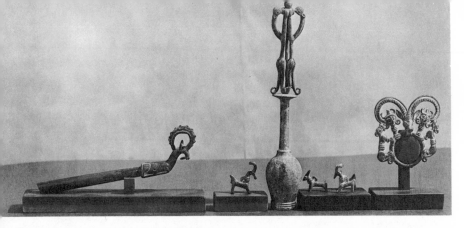

4. Luristan bronzes. Pre-Achaemenid metal work from western Iran

5. Persian ceramics. Left to right: tenth century bowl from Gorgan, twelfth century bowl from Ray, Safavid tile

6. Minor arts of the nineteenth century

7. Isfahan. The Imperial Mosque of the early seventeenth century

8. Qum. The Shrine of Fatima crowned with a gilded dome

lated to the line of the Il Khans. Thus, a certain Shaykh Hasan-i-Buzurg, "Great" Hasan, was a grandson of Arghun. In 1341 he set himself up at Tabriz as an independent ruler. On his death by assassination in 1356 he was succeeded by a son, Uways.

Uways moved from Baghdad to capture Tabriz and the surrounding region, as well as all western Iran. From Tabriz where he erected a splendid palace, he sent letters to Venetian merchants at Trebizond on the Black Sea urging them to renew their trade with Tabriz as in the days of the Il Khanid rulers.

Jalal al-Din Husayn succeeded his father in 1374 and in 1382 was defeated in battle by his brother Ghiyath al-Din Ahmad and executed. Ahmad then began a career crowded with warfare and good luck followed by ill fortune. Throughout a very wide area the situation had become very unstable due to the growing strength of the Qara-Qoyunlu to the west of Tabriz, that of the Golden Horde to the north, and the presence of Muzaffarid forces to the south. Then, in 1385 Timur swept into Azerbaijan. Ahmad fled to Baghdad and then to Syria until Timur's death. He was killed in 1410 in an unsuccessful effort to retake Tabriz, and his line survived his death by only a score of years.

Timurid Period

In the eastern reaches of Iran conditions had been unsettled for a considerable period of time. After the original Mongol invasion the lands of Transoxiana had been allotted to Chaghatay, the second son of Chingiz Khan, who there founded his own dynasty. Somewhat later the vast territory was split into two sections, Transoxiana proper and Turkestan, which carried on ceaseless warfare with each other until 1370 when Timur, then in the service of the ruler of Turkestan, managed to subdue the rival state.

Timur was born at Kish, to the south of Samarqand where his father was governor, in 1336. Timur, the name means "iron," was also known as Timur-i-Lang, "Lame Timur," because he had been lamed in one leg as a youthful mercenary. It took some years of intrigue and raids before Timur was able to establish himself as an independent sovereign at Samarqand, and it was not until 1381, when he was forty-five years old, that he set out on the first of eight campaigns to the Iranian plateau and beyond. Somewhat in the manner of the Mongol invasions, each successive campaign penetrated further to the west. His campaigns were conducted with relatively small forces, so that he frequently met with temporary reverses and even defeats. He moved on to Baghdad, and then into Syria where he took Aleppo and Damascus. His conquests, although less destructive than those of the Mongols, did work considerable havoc. One account of his exploits reads: "And as soon as they had surrendered and given up their arms, he drew the sword against them and billeted upon them all the armies of death. Then he laid the city waste, leaving in it not a tree or a wall and destroying it utterly, no mark or trace of it remaining." On the other hand, he frequently allowed the petty overlords, sultans, and amirs to retain their local power.

He also conquered much of Russian Turkestan and of India. In 1398 he returned to Samarqand, his capital, to organize the administration of his vast territories. The state which he established was essentially a Turkish kingdom but Iranian culture played the creative and dominant role, for Timur was a fervent Muslim.

A good deal is known about Timur's character, habits, and interests. According to one contemporary writer: "He was of great stature, of an extraordinarily large head and with long hair said to have been white from birth. He was of a serious and gloomy expression of countenance; an enemy to every kind of joke or jest, but especially to falsehood. He neither loved poets nor buffoons, but physicians, astronomers and lawyers. His favorite books were histories of wars and biographies of warriors. His learning was limited, being confined to the ability to read and write; three languages he could use—Turkish, Persian, and Mongolian." He seems to have loved the open air and, if the weather permitted, moved from one to another of his great gardens which girdled Samarqand. There he lived in tents or pavilions made of the most costly silks and brocades. The inherent cultural limitations of Timur did not affect his interest in the arts. He was passionately interested in architecture and gathered artisans and craftsmen at Samarqand to erect a series of structures each of which had to be higher and grander than the one before, and on his death in 1405 he was laid to rest in a towering mausoleum erected during his lifetime.

When Timur died only a grandson and a son of his sister, among a host of his possible successors, were at Samarqand. His fourth son, Shah Rukh, was at Herat, and it was he who gained control of Herat, Khorasan, and Transoxiana after nearly a year of family feud and warfare. The third son, Miran Shah, gained control of western Iran including Tabriz and Baghdad but was soon engaged in conflict with his own sons. In 1406 the Qara-Qoyunlu horde took advantage of the situation to seize Azerbaijan. In 1410 their chief Qara Yusuf took Baghdad and some years later his son acquired possession of the Isfahan region. Not until 1469 did the Aq-Qoyunlu sweep out their rivals.

In the east Shah Rukh, from his capital city Herat, reigned over all of eastern Iran until 1447. He did defeat the Qara-Qoyunlu at Tabriz and drive them into Armenia but in the end was forced to leave them in possession of the western region. Shah Rukh was one of the most cultured monarchs that Iran has ever known and made Herat the intellectual center of middle Asia. Prosperity grew under his intelligent government. Architects, painters, poets, scholars, and musicians were held in high esteem, and the artistic and literary movements of these years were to spread westward in later times and find final expression at Isfahan under the rule of Shah 'Abbas.

In his devotion to culture, Shah Rukh was ably seconded by his wife Gawhar Shad, a sister of Qara Yusuf of the Qara-Qoyunlu line, and mother of two of his eight sons, Ulugh Beg and Baysunghur.

Shah Rukh was succeeded by his eldest son Ulugh Beg, whose reign was unfortunately very brief, lasting from 1447 until 1449. He was a great patron of Persian art and literature and was much interested also in the arts of China. But his ruling passion was the study of astronomy. He built an observatory which still stands at Samarqand and established tables of calculations which were finally printed in Latin in England in 1652. Baysunghur, a brother of Shah Rukh, died from acute alcoholism but not before he had achieved lasting fame in the fields of calligraphy, music, and painting. At Herat a large group of illuminators, bookbinders, and copyists worked under his direct supervision.

Ulugh Beg was too humane, gentle, and kind for the times, and in the end was dethroned and killed by his own son, Abu al-Latif, who was himself soon assassinated. Abu Sa'id came to the throne in 1451 and succeeded in consolidating much of the area of the slowly dissolving empire, before being defeated and killed by the Aq-Qoyunlu in 1468.

With the death of Abu Sa'id the Timurid line held only Khorasan, with its seat at Herat. Within a year Husayn Bayqara, a great-grandson of Timur, replaced the initial claimant to the throne. Although his reign was anything but calm, he did create an oasis of culture at Herat and summoned artists, poets, and scholars to his court. A contemporary wrote: "The whole habitable world had not such a town as Herat had become under Sultan Husayn. . . . Khorasan, and Herat above all, was filled with learned and matchless men. Whatever work a man took up, he aimed and aspired to bring it to perfection." Among the group of luminaries at the court were men of both Persian and Turkish descent.

After the death of Husayn Bayqara in 1506 the eastern Timurid kingdom fell into the hands of a Tartar horde whose racial origin was Mongol but whose cultural tradition was Turkish. This Tartar dynasty, the Shaybanids, continued to patronize the arts at Bokhara and Samarqand but was soon brought to an end by the rising power of the Safavids.

Qara-Qoyunlu Dynasty

The Qara-Qoyunlu, literally "possessing black sheep," was a confederation of Mongol tribes, which actually carried a standard with the image of a black sheep. Earlier vassals of the Jalayirids, Qara Yusuf became independent with his capital at Tabriz, in 1389. Three times he was defeated in battle by Shah Rukh but there must have been periods of more harmonious relations, since, as noted, his sister married Shah Rukh. Qara Yusuf's son was also defeated by Shah Rukh, and then murdered by his own son. A brother, Jahan Shah, succeeded and proved to be a capable military leader who expanded his hold over much of Iran. He had a poor press in his own day: historians called him dissolute and blood-thirsty, while his habit of staying up all night to engage in revelry earned him the nickname "The Bat." His campaigns ranged as far to the east as Herat where he was enthroned in 1458. Moving west to attack the Aq-Qoyunlu, Jahan Shah lagged behind his troops to hunt and was slain by an enemy band. His son, Hasan 'Ali, held power only a few months before he was slain and the line came to an end.

Aq-Qoyunlu Dynasty

This group of Bayundur Turkmen, "possessing white sheep," first attained some fame under Qara Yuluk, "Black Leech," 'Uthman. From Trebizond he fought to expand his holdings until, having taken service with Timur in the latter's campaign into Syria, he was rewarded with the grant of Diakbakr. The succession was disputed by his sons and a grandson until another grandson, Uzun Hasan, came to the throne in 1453, the year that the Ottoman Turks captured Constantinople. The most powerful figure of the dynasty, he fought against the Qara-Qoyunlu and the Timurids, and eventually held Azerbaijan, Iraq, Fars, Khuzistan, and Kerman. He was less successful against the Ottoman forces. After his death in 1478 he was followed by a son, Ya'qub, who was "killed by an intrigue of his wife, who was not a *very* virtuous woman." What happened was that this lady offered him a drink as he left the bath with a small son. Somewhat suspicious, he asked her to taste it and then he and the boy finished it off: all were dead within hours. The strength of the dynasty waned, weakened by constant internecine strife and through attacks from tribes allied with the Safavids, until in 1501 Murad, last of the line, sought refuge with the Ottoman sultan.

Safavid Period

Early in the fourteenth century Rashid al-Din, the Il Khanid vazir, wrote a letter to Shaykh Is'haq Safi al-Din in which he expressed great respect for this revered head of a Sufi order at Ardabil to the north of Tabriz. Safi al-Din claimed to be a descendant, twenty generations removed, of the Imam Musa Kazem, and hence, still farther back, of 'Ali, son-in-law of the prophet Muhammad. After his death in 1334 at age eighty-five, his descendants became the traditional heads of this Safaviyya order. Sadr al-Din, his son and successor, was visited by Timur who freed a body of Turkish prisoners at his request: settled in the region these Turkish families became the power base of the Safi family. A grandson of the founder, Junayd, married a sister of Uzun Hasan, and his son, Haydar, married Marta, daughter of Uzun Hasan and Despina, a Christian daughter of the Greek ruler of Trebizond. Marta gave birth to three sons, one of them Isma'il. These details of genealogy are given to suggest the complex ethnic origins of the dynasty. Originally Arab, it underwent Turkish and Greek infusions. Throughout the fourteenth century the family spoke and wrote Turkish.

At the age of thirteen Isma'il started towards Ardabil from the Caspian Sea with a handful of followers. Within a year he had an army strong enough to seize Tabriz from the Aq-Qoyunlu. Crowned at Tabriz in 1500, he took the title of Shah in place of that of Shaykh of the Sufis of Ardabil. Menaced by the Ottoman Turks who were of the Sunni sect, he took the bold step of proclaiming that the Shi'a sect was the faith of his state. By so doing he employed religious unification in an attempt to achieve political unification. His decision was warmly supported by the seven strong Turkish tribes of Azerbaijan called *Qizl Bash*, or "Red Heads," after the color of their turbans. By 1510 he had taken over Iraq, Hamadan, Fars, Kerman, and Khorasan, but in 1514 his forces were defeated near Tabriz by the Ottoman ruler, Selim the Grim, and he lost his capital to the Turk. Venetian travellers who saw him at Tabriz had differing impressions: one remarked that he was as "bloodthirsty a tyrant as ever existed," while another wrote, "very much beloved . . . for his beauty and pleasing manners."

In 1524 Isma'il was buried in a chamber alongside the tomb of Shaykh Safi at Ardabil; he was succeeded by his son, Tahmasp, then ten years old. Shah Tahmasp had about ten years more in which to grow up before he lost Tabriz to Sulayman the Magnificent in 1523. He moved his capital to Qazvin, farther from the Turks, and there ruled over a court whose tone was one of refinement, sophistication, and grace. Again, contemporary observers gained different impressions of his character, while in his brief autobiography he claimed to be guided by dreams in which 'Ali and the imams appeared to him. Under their influence, at eighteen he gave up wine and closed all taverns.

In 1576 his son, Isma'il II, came from prison to the throne. Confined by his father for twenty-five years, his pent up anger burst out in an effort to exterminate his brothers and their families. Stating that "the royal tent could not be held up by old ropes," he engaged in ruthless slaughter: "He even reddened with blood his own hands, killing many of them with his own sword, saying he wanted to see whether it would cut." By happy chance a young nephew, 'Abbas Mirza—later Shah 'Abbas—escaped death. His very brief reign was ended by an overdose of opium.

Muhammad Khudabanda, a brother, succeeded. He sought to be faithful to his name, "Servant of God," by displaying mercy to most offenders, while neglecting the affairs of state and of war. The Turks were again attacking from the west, and such other enemies as the Georgians and the Uzbeks were active along the frontiers.

Shah 'Abbas came to the throne in 1588 at the age of sixteen and reigned for forty-one years. His reign was one of Iran's Golden Ages, still compared by the Persians to those of Cyrus the Great, Darius, and Khusraw I. The Frenchman, Chardin, who visited Isfahan after his death expressed the popular feeling: "When this great Prince ceased to live, Persia ceased to prosper." And even today the common man tends to assign all old mosques, bridges, and caravanserais to the time of Shah 'Abbas. The years of his reign are well-documented, particularly in the accounts of the Europeans who visited his court. Sir Anthony Sherley wrote of him: "His person then is such as well-understanding Nature would fit for the end proposed for his being, excellently well shaped, of a most well proportioned stature, strong and active; his colour somewhat inclined to a man-like blacknesse, is also more blacke by the Sunnes burning: his furniture of mind infinitely royal, wise, valiant, liberall, temperate, merciful, and an exceeding lover of Justice, embracing royally other virtures, as farre from pride and vanitie, as from all unprincely signs or acts."

The reign of Shah 'Abbas approximately coincided with those of several other rulers of renown: Elizabeth I of England, 1558-1603; Philip II of Spain, 1555-1598; Akbar of India, 1556-1605. As early as 1592 Pope Clement VII had written to Shah 'Abbas to urge an alliance against the Turks, and the European rulers were interested both in such an alliance and in trade with Iran. Their envoys to his court arrived burdened with gifts.

On coming to the throne Shah 'Abbas was faced with a most unfavorable situation. To the west the Ottoman Turks held all of Azerbaijan, while on the east the Uzbeks had moved into Khorasan, seizing Herat and Mashhad. In the years of warfare which followed the Persians gained the upper hand over these enemies.

In 1598 Shah 'Abbas celebrated No Ruz in his palace at Isfahan, and took the decision to transfer the capital there from Qazvin. Soon an imperial city was rising in an area of orchards and fields between the Zayendeh River and older Isfahan. A wide avenue, the Chahar Bagh, crossed the river on a stately bridge and headed into the new city. On either side of the avenues were gardens with splendid pavilions and between its upper end and the Maidan-i-Shah, the "Imperial Square," were strewn palaces, pavilions, workshops and storehouses. Isfahan hummed with activity in commerce and the arts and crafts. In place of an army composed of tribal forces led by tribal chiefs he created a regular, paid army which included the *Shah Sevens* or "Friends of the Shah," a force of 10,000 horsemen and 20,000 foot-soldiers. Roads, canals, and caravanserais were constructed throughout Iran. In 1622 he allied Iran with a British naval force in the Persian Gulf to drive the Portuguese from the island of Hormuz, and encouraged British and Dutch merchants to trade at Bandar 'Abbas, making it the principal center for the export of silks. He established diplomatic relations with the European countries. Kurdish tribes were moved to Khorasan to form a living barrier against the Uzbeks, and a large colony of Armenian artisans were moved from Julfa in Azerbaijan to a new Julfa across the river from Isfahan. He was tolerant of non-Muslims, zealous for public security, severe in his impartial justice, and lavish in his charitable gifts. Besides making Isfahan the architectural wonder of the world he embellished Shiraz, Ardabil, and Tabriz with fine structures. He devoted special attention to the Shrine of Imam Reza at Mashhad, probably with the intent of keeping money within the country by making the shrine attractive to pilgrims who would otherwise journey to Mecca or to the holy Shi'a shrines in Iraq. He himself made several pilgrimages, one entirely on foot from Isfahan to Mashhad, and erected new sections of the shrine complex.

Unfortunately a less noble and perhaps less sagacious side of Shah 'Abbas came to the surface in his family relations. Through jealousy and fear of being supplanted in power he put one son to death and had the eyes of two others blinded. As a result, before his death in 1629 he named his eldest grandson as his successor.

Shah Safi reigned until 1642: the Turks made inroads from the west and the Moghul forces from the east. His son, Shah 'Abbas II came to the throne before he was nine years old, and was the father of a son when not yet fourteen. He was successful in wresting the city of Qandahar back from the Moghul emperor Shah Jahan, and the other frontiers were comparatively quiet. His son, Shah Sulayman, succeeded in 1667 and reigned until 1694. He seemed to have been indifferent about external threats to the kingdom. As one man on the scene wrote: "With the king indulging in Bacchus and Venus and the officials altogether intent on making money for themselves, it is [a country] miserably abandoned."

Shah Husayn was at first an extremely pious Muslim, but turned to a life of indulgence. Filling his harem with a vast number of women, he stated that he "would spare no cost to outstrip the most voluptuous kings that ever were in the world." Conveniently, a military threat from the Afghans was neglected. In 1721 they moved west from Qandahar to take Kerman, while the main body pushed on towards Isfahan with an inconclusive engagement fought outside the capital. A number of assaults were beaten off, but during the heat of summer a grievous shortage of food afflicted Isfahan: dogs and cats were eaten in the early fall and children and adults thereafter. In October 1722 Shah Husayn donned black garments and made his way to the camp of Mir Mahmud, the Afghan leader, placing his aigrette in the latter's turban as a sign that the throne of Iran had passed to the Afghan.

Afsharid Dynasty

In 1725 Shah Mir Mahmud was put to death at Isfahan by members of his entourage, and a cousin, Mir Ashraf, took over as nominal ruler of Iran. Tahmasp, son of the deposed Shah Husayn, was in northern Iran attempting to recruit forces. He was joined by a Turkman, a member of the Afshar tribe, Hasan Quli ibn 'Ali. In his earlier life he was successively a slave, a mercenary soldier, and the leader of a strong band of robbers. A patriotic motive, hatred of the Afghan invaders, may have led him to enter the service of Tahmasp. He soon excelled as a drill master and a tactician, and Tahmasp ordered that he take his own name so that he became Tahmasp Quli Khan. Together they advanced on Isfahan and routed the forces of Ashraf who was slain in lonely flight.

Tahmasp Quli Khan then deposed his master and in 1736 was chosen shah by the dignitaries of the land and crowned as Nadir Shah. A year later he launched a campaign against India, defeating a huge Moghul army to the north of Delhi and collecting treasure valued by a contemporary at some 87,500,000 pounds sterling. Brought back safely to his capital, Mashhad, much of the treasure of gems survived and is on public display at Tehran. Successful campaigns against the Uzbeks and the Turks followed, but his increasing tyranny resulted in his death at the hands of his own bodyguard in 1747.

Three successors ruled rather briefly, but were unable to prevent the final collapse of the dynasty in the closing years of the eighteenth century.

Zand Dynasty

At Isfahan a new military leader emerged in the person of Muhammad Karim Khan, a Kurd of the Zand tribe, who brought all of Iran except for part of Khorasan under his control. Establishing his capital at Shiraz, he reigned from 1750 until 1779, a period in which the country enjoyed peace. He accepted no title other than that of *Vakil*, or "Regent," as representative of a weak Safavid descendant. A kindly and just man, he won the esteem of his people.

Karim Khan's death was followed by a bitter contest for succession among his brothers, sons, and nephews. A grandson, Lutf 'Ali, came to power in 1789: possessed of a chivalrous spirit, he displayed great perseverance in the face of misfortune.

Qajar Period

The Qajars were one of the seven Turkish tribes which had supported the rise of power of Shah Isma'il, the first Safavid ruler. Their fortunes had been at a low ebb under Nadir Shah, but after his death they became dominant in Mazanderan and made an abortive attempt to spread into southern Iran. One of them, Agha Muhammad Khan, was held at the court of Karim Khan who had defeated his father in battle. As a child he had been castrated by one of the Afsharids, and his unfortunate condition and years of restraint bred a ferocious temper. Escaping from Shiraz, he made his way to Mazanderan and united the branches of the tribe. From his new capital at Tehran, he ruled Iran from 1779, although not officially named as Shah of Iran until 1796. In the south Muhammad attacked the forces of Lutf 'Ali, of the Zand line. Lutf 'Ali sought refuge in Kerman but was finally captured and tortured to death while Muhammad punished Kerman by putting out the eyes of 20,000 of its inhabitants. By the time he himself was assassinated in 1797, Agha Muhammad Khan had gained control of the whole of Iran, including Georgia.

His nephew and successor Fath 'Ali Shah, who ruled from 1797 until 1834, was a man of quite different character. His reign ushered in a century in which Iran enjoyed comparative calm and peace while it suffered a moral and political decline. Direct contact with the European powers began under Fath 'Ali Shah with a treaty of alliance, signed in 1807, between France and Iran. Napoleon expected this treaty to open the way for a French invasion of India by land, while Iran was to receive arms and military instruction to enable her to resist the expanding strength of Czarist Russia, which had annexed Georgia in 1801. However, Napoleon soon came to terms with Russia, and the hostilities which broke out between Russia and Iran were ended in 1813 by the Treaty of Gulistan confirming Russian possession of Georgia. In 1814 Iran and Great Britain signed a treaty of defensive alliance which, although it remained in force until 1857, was never of any value to Iran.

In 1826 Iran and Russia again went to war, and initial Persian successes were followed by a series of defeats culminating in the capture of Tabriz by the Russians. The Treaty of Turkoman Chai, signed in 1828, gave Russia the Persian districts of Erivan and Nakhichevan, exacted from Iran a large indemnity, reserved military navigation on the Caspian to Russian ships, and granted capitulations in favor of Russia. A later annex to the treaty gave Russia special economic and tariff rights.

From this time until well into the twentieth century Iran was to be torn between the conflicting interests of Russia and Great Britain. Russia was embarked on a course of expansion in Asia and had visions of a warm water port on the Persian Gulf, while Great Britain was faced with the need of controlling the Persian Gulf and all land areas adjacent to India, her great colonial prize.

Muhammad Shah, the grandson of Fath 'Ali Shah, ruled from 1834 until 1848. He did his utmost to improve the internal condition of the country, abolishing the practice of torture and forbidding the importation of slaves into Iran.

During his reign Russia wooed Iranian friendship in order to have a free hand in consolidating her gains in the Caucasus and in Turkestan. Muhammad Shah, supported by Russia, made an attempt to reconquer Herat which was strongly opposed by Great Britain, who sent a British officer to organize the successful resistance of Herat.

Nasr al-Din Shah, son of Muhammad Shah, came to the throne at the age of seventeen. His immediate task was to put down revolts that broke out in six different parts of the country. Successful in this effort, he managed to reign for nearly fifty years—an assassin's bullet cut him down just four days before that anniversary. His long reign was marked by friendly relations with Russia, whose influence within Iran became firmly entrenched. In 1856 the Persian army marched into Afghanistan and took Herat. Great Britain, who had fought against the ruler of Afghanistan from 1839 until 1841, demanded the immediate evacuation of Herat; the governor general of India declared war on Iran; and British troops were landed at the head of the Persian Gulf. Russia failed to support Iran and Nasr al-Din capitulated. By the Treaty of Paris, signed in 1857, Iran withdrew from Herat and recognized the independence of Afghanistan. The treaty also granted capitulations and special commercial privileges to Great Britain.

The rivalry between Russia and Great Britain in the Iranian theatre now took the form of economic penetration. Since the growing industrialization of the West demanded both access to raw materials and new markets for manufactured products, certain distant countries were marked for economic penetration, necessitating some degree of political intervention. Within Iran this policy found one application in a struggle for concessions. In 1872 a British banker, Baron Reuter, obtained an amazing concession from Nasir al-Din. Detailed in more than twenty articles, it gave Great Britain the right to construct railways and street car lines, to exploit minerals and oil for a period of seventy years, and to manage the custom service for twenty-four years. When Nasir al-Din made his first trip to Europe in the following year he was very coldly received in Russia and upon his return canceled the concession. However, in 1889 he placated Reuter with a concession for the creation of the Imperial Bank of Persia. In 1890 a British concern was given a tobacco monopoly, but the clerical leaders of the country supported a wave of general indignation by formally forbidding the use of tobacco within Iran until the monopoly was canceled. After 1863 Great Britain was also active in promoting the erection of telegraph lines across western Iran.

Russia was not idle. In 1879 Nasir al-Din agreed to the creation of a brigade of Persian Cossacks patterned on the Russian model and instructed and commanded by Russian officers, and forces were soon established at Tehran and other northern towns. The Discount Bank of Persia, a Russian institution, was opened at Tehran in 1891. In 1888 a Russian subject gained a comprehensive concession covering fishing rights in the Caspian. Meanwhile Russia was very active within the ancient domains of Iran: after 1865 her armies took Tashkent, Samarqand, Bokhara, and Khiva, and in 1882 Iran signed the Treaty of Akhal which gave Russia possession of the important city of Merv.

The American Legation at Tehran was opened in 1882, and between 1855 and 1900 at least fifteen foreign countries gained capitulation rights for their subjects residing in Iran.

Nasir al-Din did the best he could for his country, but circumstances were too strong to be effectively countered. He made three lengthy tours of Europe and kept diaries which were printed in Persian. They display an avid curiosity about all he saw and heard. Military matters were a major interest; with the need for the modernization of Iran's army in mind he visited munitions plants, cannon foundries, and powder plants, and looked into the possibility of bringing military advisers to Iran. He became convinced that Iran needed only to adopt Western skills and methods to take her place in the modern world. He soon learned that this was not the case, for the concessions granted to foreign powers and subjects brought little advantage to Iran, while Tehran was crowded with adventurous strangers who hoped to make their fortunes at the expense of the gullible Persians. The Shah was, however, by no means a naive figure. While in Europe he had been unfavorably impressed by the military reviews and continual preparations for war made by each nation. Although the Shah made a serious attempt to improve the systems of justice and of public administration, his efforts were not crowned with lasting success and the country came increasingly under the influence of the clergy.

His son, Muzaffar al-Din, reigned from 1896 until 1907. After a youth passed in idleness and the pursuit of pleasure at Tabriz, as sovereign he showed no more force or true concern for the affairs of state. His trips to Europe cost tremendous sums, often leaving the treasury almost without funds. Nobles and courtiers amassed fortunes while public officials failed to receive their salaries; a few landlords acquired tremendous holdings while the peasants were squeezed dry. Irrigation works fell into ruin, and the desert encroached on villages and fields.

At length the need for funds led Iran to secure from Russia, in 1900, a loan of 22,000,000 roubles at five per cent interest. Under the terms of the loan, part of the sum was to be used to pay off all other debts to foreigners and until the loan was repaid Iran could not borrow elsewhere without Russian consent. Payment was guaranteed by Iranian customs receipts. At the same time Iran renewed an earlier secret agreement not to grant any railway concession to foreigners without Russia's consent. Russia also began the construction of carriage roads in the north of Iran.

When the expenses of the Shah's visit to Europe in 1900 had been met, the debt to the Imperial Bank paid up, and other obligations settled, only 6,000,000 roubles of the Russian loan remained. Russia promptly made another loan of 10,000,000 roubles and Muzaffar al-Din as promptly left again for Europe. In 1901 Russia and Iran signed a customs agreement which provided for low tariffs on goods normally supplied by Russia and higher tariffs on goods furnished by other countries.

At last the time was ripe for change. Agitation for a Constitution seemed to spring up full blown but actually it had been strongly rooted in the contact of the younger, educated class with the liberal thought of the West. The merchants and many of the clergy and nobles also supported the Constitutional Movement, some in the expectation of financial or political advantage. In July 1906, the merchants and other dwellers of Tehran to the number of 10,000 flocked to the grounds of the British Legation where they took *bast,* or "sanctuary," and were safe from arrest, while the clergy left the city. Economic and public activity came to a standstill, and pressures on the Shah mounted. In August 1906 he granted a form of a Constitution which provided for the election of a National Consultative Assembly, and the very ill ruler attended its inauguration in October 1906. This body then drafted a Constitution which was signed on 30 December 1906 by the Shah, and by his son, Muhammad 'Ali Shah.

Muhammad 'Ali Shah succeeded in January 1907. W. Morgan Shuster has described him as "perhaps the most perverted, cowardly, and vice-ridden monster that had disgraced the throne of Persia in many generations." Counseled by evil advisors, he expected to be able to take advantage of dissensions within the ranks of the Constitutionalists to restore absolute royal power. In fact, the merchants were failing to supply funds to support the National Assembly, and influential members of the clergy turned against the secular provisions of the Constitution.

In September 1907 came the announcement of an Anglo-Russian Agreement. Great Britain had supported the Constitutionalists, who had demanded an end to Russian intrigue and influence at the Persian court. Russia had just emerged from a losing war with Japan. Great Britain sensed a new menace in a German plan to build a railroad across the Near East to the Persian Gulf, and proposed an agreement with Russia which would also serve as a means of defense against German ambitions. The 1907 Agreement contained provisions relating to Iran, Afghanistan, and Tibet. With regard to Iran, the contracting parties agreed to respect her integrity and independence and then proceeded to divide the country into zones of influence. Britain was to refrain from seeking any public or private interests in the Russian zone, which took in the entire northern part of the country and included the towns of Tabriz, Rasht, Tehran, Mashhad, and Isfahan. The zone reserved for English interests was much smaller, covering the southeastern corner of Iran. The area between the two zones, although not specifically defined in the agreement, was to be neutral. In actual fact it became a field of British activity and a barrier on the road to India.

Meanwhile the Shah, encouraged by Russian assurances of support, was ready to move against the Constitutionalists. In June 1908 the Persian Cossack brigade, commanded by the Russian Colonel Liakhoff, bombarded the Parliament building with several casualties, and the Shah proclaimed the dissolution of Parliament. The public response was swift and decisive. At Tabriz the revolutionists held the city until a Russian force entered and with considerable violence suppressed the so-called disorders. Revolutionary forces recruited at Rasht and Isfahan, the Isfahan force consisting of 5,000 Bakhtiaris led by one of their tribal chiefs, marched toward Tehran. The Persian Cossack brigade was defeated outside of Tehran and in July 1909 the liberal troops entered the city. Muhammad 'Ali Shah first sought refuge in the Russian Legation and then fled to Russia, and the reinstated Parliament named his eleven-year-old son, Ahmad Shah, ruler of Iran. Russian troops remained in northern Iran.

The victorious revolutionaries soon fell into disagreement, and progress was slow until an American, W. Morgan Shuster, was engaged as Treasurer-General of Iran. With several American assistants, Shuster arrived in Tehran in 1911 and in a very short time had made considerable headway with a reorganization of the financial system, while his energy and obvious devotion to the best interests of the country won the people's hearts. Russia strenuously opposed his work, and in November 1911 presented Iran with an ultimatum demanding, among other things, that Shuster be dismissed. To put teeth into it Russian troops advanced as far as Qazvin, slaughtered many of the liberals at Tabriz, and at Mashhad bombarded the shrine of Iman Reza. The ultimatum was rejected by Parliament but accepted by the Cabinet, and Shuster left Iran to write his classic, *The Strangling of Persia*.

After the outbreak of World War I Iran declared her neutrality, but Tehran became a hotbed for intrigues of Russian, British, and German diplomats and agents. In the northwest of the country Turkey and Russia maneuvered for position, and a Turkish force advanced half the distance from Baghdad to Tehran before being defeated by the Russians. In 1916 Major Percy Sykes came from India to Bandar 'Abbas and recruited a force called the "South Persia Rifles," which gained control of the entire southern section of Iran and reached a strength of 5,000 men. In the vicinity of Shiraz a remarkable German agent named Wassmuss stirred up the tribes, while special German missions tried to cross Iran to win Afghanistan to their cause.

After the war Iran's claims were rejected by the Peace Conference, and she was confronted in 1919 with a treaty proposal drawn up by the British. Under its terms Great Britain once more pledged to respect the integrity and independence of Persia, and agreed to furnish expert administrative advisors, to be paid by the Persian government, as well as military advisors and arms and equipment, also at Persia's expense. Persia would receive a loan, Great Britain would construct roads and railroads and the existing customs agreements would be restudied. The treaty would have placed Iran under complete British domination. Shah and Cabinet seemed ready to accept it, but popular feeling, encouraged by an American diplomatic protest, ran so high that it was never ratified in spite of strong British pressure. The entire country was now in a state of near anarchy. Bolshevik troops were in force along the Caspian littoral, and there was fighting between Soviet troops and a British expeditionary force.

Suddenly, in 1921, Iran and Soviet Russia concluded a treaty of friendship which represented a complete reversal of the Czarist policy toward Iran. Soviet Russia declared that all treaties and agreements formerly in effect between Iran and Russia were ended, as well as all agreements between Russia and a third power which were harmful to the best interests of Iran. She canceled all outstanding debts of Iran, voided all Russian concessions, and turned over to Iran such Russian assets on Iranian soil as the Discount Bank of Persia, the railroad from Julfa to Tabriz, the port of Enzeli, and roads and telegraph lines. She also denounced capitulations and gave Iran equal navigation rights on the Caspian Sea. One section of the treaty reflected the fears of the new Soviet government: neither state was to permit activity within its territory by groups or organizations which had designs against the other state. Each was to keep out of its country troops of a third power which threatened the security of the other. But if a third power should create such a threat within Iran, or attempt to turn Iran into a military base for action against Russia, and if Iran herself should be unable to cope with this danger, Russia reserved the right to send her troops into Iran in self-defense.

III. PATTERNS OF IRANIAN CULTURE AND SOCIETY

The Continuity of History

IRAN remains headed by a *shahinshah*, the "king of kings," and probably no other country has borne a single name as long as has Iran, nor been ruled by as many monarchs bearing the same title.

Over the long centuries since the rise of the Achaemenid empire some thirty-three dynasties ruled Iran. A number were represented by two or more branches which held sway over various areas, so that the total number of dynasties is at least forty-six, and in those lines there were some 446 rulers. These dynasties include a number aptly described by a Persian historian as those of the kings of forgotten names.

As would be expected, the rulers were conscious of their fleeting mortality, and took steps intended to preserve their memories. These actions include the patronage of scholars, artists, poets, and men of science, and devotion to pious deeds. In the Islamic centuries expressions of piety included the founding of hospitals and religious schools, the construction of mosques and shrines, the feeding of the poor, and the extermination of sects held to be heretical.

As the historical review has shown, the times were nearly always out of joint. Pious deeds were often overshadowed by acts of savagery, regarded as vital to the preservation of the throne.

On many occasions the role and qualities of the shah were defined. In all periods he was the defender and propagator of the faith: in Islamic times he was the shadow of God upon the earth. In the *Shah nama* Firdawsi stated that Allah's order is the shah's order. Allah wanted man to obey His orders, those of the prophet and those of the shah, since there was no difference between them. In his *Siyasat nama* Nizam al-Mulk wrote, "in all ages the Great God chooses one from among mankind and bestows the arts of ruling upon him. He makes him the guardian of the prosperity and peace of his people; he closes the door of corruption and evil to him; and his might enables the people to live in justice." In another work the qualities of the shah were said to include wisdom, avoidance of hasty action, devotion to truth and honesty, and sternness with mercy.

In a word, the ideal shah possessed in unrivaled quantity the most valued characteristics of the society. They were not, however, the qualities of the ideal man, for the shah could exercise qualities which were beyond the attainment of others. For example, for a long, long time the individual in Iran has desired, usually vainly, absolute justice. Writers praised the devotion of rulers to justice; a Sasanian ruler was honored as Anushirvan the Just and in later centuries many folk tales told of his absolute justice. In spite of consistent injustice, the ideal of impartial justice remained alive, and there was movement towards this goal. One lesson of the history of Iran is that man's inhumanity to man came under critical examination, and was not accepted as natural and inevitable.

To the Persians, history is not a tale of the long-distant past: the time scale is compressed. The greatest periods of history surpassed the more recent ones in which prosperity and armed strength declined. While the Persians take pride in the achievements of Shah 'Abbas and Nadir Shah, their real admiration is reserved for the Achaemenid and Sasanian periods. Viewed, as it were, without the perspective of time, they may have occurred just yesterday. Of course, this concentration on dynasties of Iranian blood is related to the fact that few later dynasties sprang from Persian stock.

Historians long emphasized the role of the ruler in the life of the country: his battles, his deeds, a chronicle of his times. In fact, there has been almost no other kind of historical writing until modern times. Naturally, then, the emphasis has been upon the institution of the monarchy, and it has been the single institution to endure from remote times. Even today the village people regard the ruler as the *khuda-yi kuchik*, or the "small God," in contrast to Allah, the *khuda-yi bozorg*, or "great God."

The Continuity of Culture

Iran's civilization has its own distinctive character. One force which moulded it—a static one—was the geographical location of Iran as a land bridge between East and West, offering positive advantages to the growth of civilization and culture while at the same time inviting disaster. On the one hand, the lofty plateau was a point of convergence for travelers, traders, and men of learning and for intellectual and artistic currents, and during each successive period Iran was in rewarding contact with powerful civilizations which arose, in their turn, to the east or to the west of the plateau. On the other hand, the plateau region was from earliest times an attractive corridor for the westward sweep of migratory groups as well as a rich goal for booty-seeking conquerors. Invasions and conquests frequently devastated the country beyond apparent possibility of recovery, but after each bitter trial its culture re-emerged.

Iran met the challenge of its geographic position through its capacity for assimilation. This capacity for absorbing foreign blood and adapting foreign influences was due both to the topography of the land and to the vitality of its inhabitants. Peoples who entered Iran in mass migrations or as a victorious army tended to settle in an environment they found more attractive than their places of origin—the barren steppes to the northeast, the hot lowlands to the southwest, or the desert wastes of Arabia.

In the course of long centuries the total number of the newcomers far exceeded that of the original inhabitants of the plateau, but they were all absorbed by and became a part of Iranian civilization, a conversion favored by the low cultural development of the areas in which the newcomers usually originated, and by the force and attractiveness of the Iranian cultural modes. Probably the most striking example of this absorption is that of the Mongols, who entered the plateau as barbarians reveling in slaughter and destruction and after two generations of settled existence became fervent admirers and advocates of every aspect of Iranian life.

Iran was always ready to receive and to recombine foreign ideas, influences, and specific artistic forms, a tendency which was apparent in prehistoric times and was specifically underlined by Herodotus when he wrote of the willingness of the Achaemenids to adopt foreign customs. The history of art and architecture is full of examples of this assimilative capacity. However, in the fields of artistic expression, of social organization, of religion, and of philosophy, elements admired merely because they were foreign and new were never slavishly copied. Instead, they were always restudied, reworked, and reexpressed in a characteristic fashion.

The history of Iranian culture exhibits a remarkable persistence and continuity which in itself is certainly not peculiar to Iran, but the specific enduring elements which produced it and the manner in which they were expressed are distinctive: pride in the past; type of social structure; character of artistic expression; the search for the meaning and purpose of life; and the Iranian outlook and attitude toward the surrounding world.

We may speak of Iran's pride in the past and of its continued awareness of the value of its cultural heritage, but we cannot speak of the inhabitants of the plateau as "patriotic" Iranians; most of them never thought of themselves as "Iranians" or as loyal supporters of some such pregnant symbol as "the homeland" or "the divine ruler." The closest approach to such a state of mind came in the early Safavid period when religious unity in Shi'ism, military unity against foreign armies, cultural unity in a period of intellectual activity, and personal loyalty to the Safavid dynasty served to establish a common bond throughout the empire. Continuity with the past was expressed in a variety of ways. The Parthians thought of themselves as the political heirs of the Achaemenids, and the Sasanians pursued a policy parallel to that of the Achaemenids in establishing a national state with its national religion. 'Adud al-Dawla, the Buyid ruler, identified himself with this past by removing four stone doorways from the palace of Darius at Persepolis and re-erecting them in his own palace near Shiraz. And one of the royal titles of Shah 'Abbas was "the splendor of Darius."

The ancient festivals of pre-Islamic Iran survived. That of *No Ruz*, New Year's Day, so celebrated in Achaemenid and Sasanian times, was never neglected. In 1595, during the reign of Shah 'Abbas, Qazvin was brilliantly illuminated for ten nights. The public was entertained and fed, while the ruler played polo, made appointments and promotions, and distributed robes of honor. He also observed the Sasanian festival first called *abhrizaghan*, or "scattering water," which marked the appearance of heavy rains after a prolonged drought. It retained the same name in the form of *ab pakhsh*, and at Isfahan the people went into the river and threw water over each other. In Gilan the ruler took part in an ancient festival, that of *Panja*, or "five," in honor of the first five days of summer.

Firdawsi's *Shah nama* was a source of pride for kings and commoners in its evocation of the legendary past of Iran. Illiterate people learned long passages by hearing constant repetition. At the court of Shah 'Abbas were several *Shah nama-khwans*, or "reciters of the *Shah nama*," and even such a ruler as Muhammad 'Ali Shah, the Qajar, had the epic read to him by his secretary.

Before moving on to other reflections of the continuity of culture, two introductory remarks should be made. First, the break between pre-Islamic and Islamic Iran was not as drastic as one would expect; and, second, science, literature, and the arts, seemed to have flourished during periods of turmoil and adversity, and to have declined in those of peace.

Artistic expression in Iran had its own marked and continuous character. Iranian art has always been decorative and, normally, non-representational. It is, of course, unfair to allow western prejudice in favor of representational art to result in an unfavorable judgment of decorative art. Decorative and abstract art carries its own forceful meaning through the use of symbols rather than of pictures, and Iran early developed a vocabulary of forms and patterns which had permanent meaning and validity for its people. The earlier works of art established standard compositions which endured with little change for centuries, although they were sometimes reworked as new artistic media came into popularity. Iranian art was always characterized by precision, clarity, and lucidity; for example, in ceramic production interest in a display of technical skill resulted in great variety but never worked, as it did in the Far East, toward the creation of shapes which deny the quality of clay. Iranian ornament, however colorful and elaborate it might be, was always based upon a clearly visible foundation pattern and never became involved in the restless and confused exuberance common to post-Renaissance ornament in Europe and to later Indian decoration.

Iranian art shows a steady stylistic development of the type common to Western art, but it went forward at a much slower rate and was not marked by abrupt changes in style. There was no real break between pre-Islamic and Islamic art: familiar decorative forms were continued, and the basic plans and methods of construction common to Sasanian fire temples and palaces reappeared in the Muslim monuments of the country.

Islam did bring with it a prohibition against the representation of living forms, a characteristic feature of Semitic religions. Although this prohibition was not always observed in Iran, its existence favored the natural predilection for decorative detail. Architectural monuments were clad with floral and geometric designs, executed in plaster, brickwork, and glazed tiles. Plan types and methods of construction remained the same for centuries, for the Persian builders were seldom intrigued by the unusual, and never felt the urge for experimentation common to the West.

From very early times craftsmen in the bazaars made and sold their products in the same shops. By extension the arts and architecture of Iran were produced by craftsmen and artisans in villages and towns who were part of the continuing currents of culture. They belonged to families which were active over many generations: their horizons were limited, but their devotion to tradition was unlimited.

Religion

The prophet Zarathushtra, known later to the Greeks as Zoroaster, is believed to have been active in eastern Iran shortly after 600 B.C. As a priest of the old Aryan faith, he severed these ties and proclaimed his mission. Our knowledge of his teachings comes from the *Avesta*, a fragmentary collection of hymns, legal codes, and rituals of various dates. It was not written down for many centuries; in fact, the oldest surviving manuscript dates from the thirteenth century A.D. Claiming to be the chosen one of the god Ahuramazda, Zara-

thushtra preached the eternal conflict between Good and
Evil, between Truth and the Lie; and he advocated Good
Thoughts, Good Words, and Good Deeds. Ahuramazda was
opposed by Ahriman, the personification of evil and darkness.

Zarathushtra instructed his followers to give up the cults
of the *daivas*, or "demons," and to abstain from sacrificing
cattle and from using *haoma*. The Aryans, whose way of life
was reflected in the *Avesta*, were sedentary cattle breeders and
farmers who held cattle in very high respect. They worshiped fire and water, and considered the pollution of the
earth a great evil.

The religious practices of the Achaemenids reflect Zoroastrianism, although Zarathushtra is never named in their inscriptions. In one inscription Darius states that Ahuramazda
is the god of the Aryans and in another proclaims, "A great
god is Ahuramazda, who created the earth, who created yon-
der sky, who created happiness for man, who made Darius
king. . . ." Relief carvings depict Ahuramazda as a bearded
figure within a winged disk. Sometimes he hovers over a fire
altar at which a king worships. However, the Achaemenids
sacrificed animals, with these sacrifices presided over by the
Magi, a clan of the Medes which became the hereditary priest-
hood. Also, they consumed haoma, a ritual liquor. In addition, the lesser Aryan divinities, Anahita and Mithra, were
worshiped.

While the worship of Ahuramazda continued in Parthian
times, the religion, then called Mazdaism, reached its high
point in the Sasanian period. There was a numerous clergy
arranged according to rank, with two Magi resident in each
village. For many years Kartir was the head of this clergy
and in four inscriptions relates his efforts to spread the faith,
to establish an orthodox form, and to put down other religions. Sacred fires burned in fire temples and in the open
air; there were national fires, fires established by the rulers,
so-called *varahan* fires, and village fires. Ritual was increasingly emphasized, while Anahita and Mithra continued in
favor.

Other religions made some headway within the empire.
Mani, described as a reformer who emphasized spiritual

needs, and the alleged author of the Book of Shapur, preached during the reign of Shapur I and before many years his message reached the Roman Empire. His teaching was tolerated by Shapur I, but about A.D. 274 Kartir accused Mani of heresy and presided at the trial in which Mani was condemned to death. However, Manichaeism survived in eastern Iran and central Asia.

Mazdak, a native of Khorasan, appeared as the dualist reformer of Zoroastrianism and advocated such specific precepts as non-violence, vegetarianism, and communism. The ruler Kavad at first supported the new sect as a possible lever against the power of the clergy and the nobles, but the apostle and his followers were massacred in A.D. 528. However, the Mazdak sect persisted into Islamic times. Christians were frequently, often violently, persecuted in the first centuries of the Sansanian period, primarily because of the identification of the religion with the rival Roman Empire. After the end of the fifth century, when the Christians within the empire were members of the eastern Nestorian Church, they were treated with increased tolerance.

As noted earlier, in the seventh century A.D. invading Arab armies brought a new religion to Iran. *Islam*, as used in the Qoran dictated to the Prophet Muhammad, means "submission to the will of God." Its five obligations are: the confession of faith expressed in the formula "There is no God but Allah and Muhammad is His Prophet"; the five daily ritual prayers; the fast from sunrise to sunset during the month of Ramadan; required and voluntary alms; and the pilgrimage to Mecca.

The sons of Muhammad died in childhood. His daughter Fatima, who married 'Ali, was the only one of his daughters to give him male heirs. The two sons of 'Ali and Fatima were named Hasan and Husayn.

Upon the death of Muhammad the question of a Caliph, or "Successor," arose at once, and the choice fell upon Abu Bakr, who held the post but two years, from 632 until 634. Then came 'Umar, who was assassinated in 642, and after him the aged 'Uthman, another son-in-law of Muhammad. Dissatisfaction against 'Uthman arose on many sides, and a respected companion of Muhammad and other influential people began to preach in favor of 'Ali and of the rights of his family to the Caliphate, or headship of the religion. In 656 'Uthman was killed and 'Ali was chosen as Caliph. Moawiya, governor of Syria and cousin of 'Uthman, refused allegiance and fighting broke out, and in 661, 'Ali was assassinated by a member of still another hostile group.

After his death the Shi'ites, or the "partisans" of the family of 'Ali, chose his son Hasan as his successor. However, the power was held by Moawiya, who moved the capital from Medina in Arabia to Damascus, and whose line was known as the Umayyads. Before his death Moawiya named his son Yazid as his successor but Husayn, the younger brother of Hasan, was urged to press his rightful claim to the Caliphate. He and his small band of faithful followers were surrounded on the plain of Kerbala in Iraq, and after a period of ten days they were all put to death.

In 749 the Umayyads gave way to the 'Abbasids whose line was descended from the uncle of Muhammad. The 'Abbasid Caliphs settled at Baghdad and held temporal and spiritual power as heads of the Sunni sect of Islam until the line was wiped out by the Mongols in 1258.

While the Iranians became ardent Muslims, their established traditions and culture had an impact on the faith which one scholar has described as the Persian conquest of Islam. For example, they claimed that Husayn had married a daughter, Shahrbanu, of the last ruler of the Sasanian empire, and they developed special dogma, such as that of the sinlessness of the family of 'Ali, and of Husayn's sacrifice in atonement for the sins of mankind.

The Shi'a sect gained popularity in Iran because of its nationalistic connotations, as differentiating the Persians from the Arabs. The Buyid rulers were Shi'as, but the foreign dynasties which ruled Iran followed the more orthodox Sunni sect and actively persecuted the Shi'as. As a result, the Shi'as were authorized to resort to *taqiya*, or the concealment of their belief by dissimulation.

According to the Shi'as, the true successors to the spiritual leadership of Islam continued through the line of 'Ali in the series of the twelve Imams. Each Imam of the line was considered to have divine infallibility, each could work miracles, each named his own successor, and each met with death by violence. The Twelfth Imam, known as the Hidden Imam, disappeared into a cave while still a young man, and it is believed that he will eventually return as the Mahdi when the world approaches its end. Divisions arose within the Shi'a sect over the question of the line of the Imams. The Zaid Shi'ites of Arabia place their faith only in the first five, while the Karmathians and Isma'ilis recognize only the first seven.

The Shi'ites of Iran believe in the entire line of the Imams, and follow the teachings of J'afar, the sixth Imam, who was greatly respected as an authority on law and tradition. Shi'ism came into its own in Iran with the Safavid rulers whose ancestor, Shaykh Safi al-Din, claimed descent from the seventh Imam, and has been the official religion of Iran since about 1500. The Constitution designates the Jafarite sect as the religion of the country, and provides that the article in which the faith is proclaimed is to remain in effect until the appearance of the Mahdi, the Imam of the Time. Only he will be able to bring about the triumph of good over evil, and of justice over inequity. Until then, the material world and all its creatures represent a transitory state in which evil holds the upper hand. The Shi'a doctors of religious law are held to be inspired by the Hidden Imam, and their statements are granted much greater authority than those of the Sunni theologians. The Imams are worshipped as martyrs, and the Shi'ites make pilgrimages not only to Mecca but also to the shrines of 'Ali at Nejef and of Husayn at Kerbala, both in Iraq; of 'Ali Reza, the eighth Imam, at Mashhad; and of Fatima, the sister of Imam Reza, at Qum. There are also many magnificent shrines of saintly Muslims, as well as innumerable tombs of lesser figures, the *imamzadehs*, or the alleged "descendants of the Imams."

Numerous religious holidays revolve through the seasons, according to the lunar calendar of Islam, and the Shi'as mark the tragedy of Kerbala with sermons, processions of flagellants, and the type of drama called *ta'ziyeh*. These tragic plays are grouped into three cycles—the state of affairs before the battle at Kerbala, the battle itself, and the aftermath of Kerbala.

The authority of Islam long pervaded public and private life. A very loosely knit religious hierarchy was in touch with all the population: at the levels of the masses through the *mullas*, and at the court of the shahs through the *mujtahids* (a restricted group of canonists) and the *ayatullahs* (the most learned and respected religious figures). Until recent times the *shar'ia*, the religious law, was the sole legal code. Charity, a basic tenet of Islam, found a formal expression in the *vaqf*, or religious endowment: mosques, schools, and hospitals were built and the income of donated lands assigned to the upkeep of these institutions, as well as to other worthy purposes. For many centuries the *mullas* taught, in return for a very small sum of money from each parent, in schools called *maktab*, where the children memorized the Qoran by chanting its verses in unison and learned to read and write Persian and do simple arithmetic. There were also many religious colleges, something like the western theological seminaries, where advanced students, gathered around men renowned for their learning, worked at such subjects as the interpretation of the Qoran, religious law, and religious philosophy. There were no formal examinations. No place for girls was provided in this system of education.

Over the centuries efforts were made to bring together the moral principles and duties of Muslims into an integrated system of *akhlaq*, or ethics, which may be defined as a practical philosophy. Many such books of *akhlaq* were written. There was, however, a polarity of ethical behavior between these books, and the actual practice of morality—advocated by the poet Sa'di in terms of expediency and common sense.

There was also a polarity in expressions of Islam. The *sufis*, or devotees of mysticism, sought a transcendental interpretation of the transitory phenomena of existence. According to their conviction, mortal flesh suffers and the soul longs for union with the one reality, its creator, and for escape from this realm of conflict between good and evil to that of identification with absolute good. In contrast, poets moralized on

[84]

9. Muhammad Reza Pahlavi Shahinshah Aryamehr and the Shahbanu, Empress Farah

10. Bejeweled ewer and basin from the magnificent collection of the Crown Jewels of Iran

11. Isfahan. The dome chamber, covered with glowing faience mosaic, of the Mosque of Shaykh Lutfullah, erected in the early seventeenth century

the uncertainty of human affairs. Rather than devoting attention in preparing for a better life to come and union with the creator, as the sufis taught, he should seek to enjoy his fleeting existence. The earlier Western scholars who saw, falsely, Islam pervaded by fatalism were followed by those who found fatalism in this poetry. What these critics seem to have failed to realize was that this poetry is permeated with concern for man's fate. In summary, with Iran torn throughout many centuries by devastating warfare and by man's inhumanity to man, the Persian had the choice of denying the validity of the material world, or of coming to terms with it.

While the role of Islam in modern life will be considered in a following section, it should be stated that ninety-eight percent of the people are Muslim, and ninety percent of that number are Shi'as.

Language and Literature

The Achaemenids, that union of the Medes and the Persians, spoke the vernacular language which linguists call Old Persian: it was a member of the Iranian branch of the Indo-Iranian, or Aryan, group, one of the main divisions of the Indo-European family of languages. The carved inscriptions of the rulers were usually trilingual, executed in cuneiform signs in Old Persian, Elamite, and Akkadian (Babylonian). Court records were written in Elamite and Aramaic. Properly speaking, there is no Achaemenid literature. The inscriptions carved for the rulers total about 5,800 words, and, with the notable exceptions of the long inscription of Darius at Behistun and that on his tomb, are largely repetitious statements of kingship and relationship, of piety, and of lists of the lands held by the Achaemenids.

The language of the Parthian, or Arsacid, period is called Arsacid Pahlavi by scholars, with Pahlavi the name generally applied to so-called Middle Persian in use from 300 B.C. until A.D. 900. The language was written heterographically with Aramaic letters and even words. What this meant was that the same spelling could represent different sounds and even different words. One scholar may read an inscription of the period in Aramaic letters as if it were in the Aramaic language, while another may read it as if it was in Pahlavi: their translations would be completely different.

The language of the Sasanian period continued to be Middle Persian written in the awkward and intricate Pahlavi script, but the language lost a good deal of its grammatical complexity and began to approach the final form of New Persian, or *farsi*. Among the secular writings were the Book of Great Deeds, the Book of Rank, the lost Book of the Kings, the Lands of Iran, and translations such as that of the fables of Kalila and Dimna. Certain lost works exist in later Arabic translations. In this period the Avesta was written down several times, and Shapur I recorded it in a special script which is known to us as *avestic*.

New Persian was well established as a literary language by A.D. 900, written in the Arabic alphabet and script. In the region of Fars it took the form called *farsi*, the name still applied to the Persian of modern Iran. In the eastern Iranian plateau it took another form, *dari*, or "of the court." More archaic, less ornate, and less penetrated by Arabic, it lost its literary importance in the eleventh century. In this same region appeared an anti-Arab movement called *shu'ubiyya*, which may have had two goals: to perpetuate old Iranian traditions and to stress the vital role of Iran within the newer world of Islam. Thus, there was emphasis on legends of the rulers and warriors of the pre-Islamic centuries.

At about this same time Persians who had become proficient in Arabic wrote major works in that language. Ibn Khaldun, an Arab from North Africa whose brilliant work, the *Muqaddimah*, or "Introduction to History," was written near the end of the fourteenth century, had this to say: "All of the founders of grammar were of non-Arab [Persian] descent. . . . Most of the *hadith* scholars who preserved traditions for the Muslims were also non-Arabs [Persians], or Persian in language and upbringing. . . . Only the non-Arabs [Persians] engaged in the task of preserving knowledge and writing systematic works. Thus, the truth of the following statement by the Prophet becomes apparent: 'If scholarship hung suspended in the highest parts of heaven, the people of Fars would reach and take it.' "

This scholarly activity had the effect of impregnating New Persian with Arabic. There was a flow of Arabic words, and also of grammar, rhetoric, and prosody. Persian poetry was invaded by the elaborate figures of speech and thought common to Arabic poetry. This penetration was not, however, disastrous for the future of Persian literature. On the one hand the interaction of Aryan (Iranian) and Semitic (Arabic) cultures was a positive force, and, on the other, the intrusion of Arabic provided Persian with an extraordinarily copious vocabulary. Then, Turkish words intruded into Persian during the Il Khanid Period, and more seriously in the early Safavid period.

Persian literature is amazingly rich, extensive, and varied, not inferior in these respects to any other national literature. The subject matter includes universal, general, and regional history; geography; accounts of travel; encyclopedias; anthologies; fables; anecdotes; essays; ethics; manuals of conduct and practical wisdom; grammar; philology; mysticism; theology, including exegesis and tradition; jurisprudence; medicine; astronomy, including astrology; chemistry, including alchemy; and biography. There is little autobiography, and the novel appeared only in recent times, as the result of the influence of European literature.

Then, there is poetry. By temperament the Persians are natural poets; poetry rather than blood may run in their veins. The love of poetry was so strong that it affected prose, which often displays rhythm and rhyme. Poetry is classified as epic or lyric, with the latter term covering a wide range of forms. Epic poetry employs the *mathnavi*, that of rhyming distiches. Most popular in all periods was the *ghazal*, the ode or lyric, with from four to sixteen couplets, in which the poet included his *takhallus*, or "pen name," in the final hemistich. The *qasida*, the ode or elegy, was longer than the ghazal, and maintained, as did the ghazal, an identical rhyme throughout. The *rub'ai*, from the Arabic word for "four," with its Persian plural *ruba'iyat*, was, of course, the quatrain.

[87]

The figures of speech are far too many to be cited in full. Alliteration was of great importance, as was metonymy. Figures of thought are nearly as numerous. Hyperbole is the most common, while another much favored is ambiguity. Indeed, ambiguity is so general that it has been said that it is almost impossible to translate a line of Persian poetry into another language without neglecting at least one of its possible meanings.

The collected works of a poet are known as his *divan*, or his *kulliyat*. Many major monuments of literature have survived only as fragments cited in later anthologies and biographies of poets. All the literature was, of course, in the form of manuscripts which were copied and recopied over the centuries, many of them illustrated with miniature paintings.

Renowned among the early writers in dari was Rudaki (Abu 'Abdullah Ja'far ibn Muhammad Rudaki), probably from the village of Rudak to the east of Samarqand. He wrote at the Samanid court in Bukhara, and died about 940. A number of his qasidas are preserved in much later anthologies. Daqiqi (Abu'l Mansur Muhammad ibn Ahmad Daqiqi) was called to the same court about 950 and commissioned by the ruler to compose an epic about pre-Islamic Iran. Before his death he had completed 1,000 couplets on the life and times of Zoroaster, which were incorporated into Firdawsi's epic.

Mahmud of Ghazni summoned to his court men of learning and few dared decline the invitation. A contemporary wrote that the court sheltered 400 poets. Preeminent among them was Abul Qasim Hasan ibn Ahmad Balkhi, called Unsuri, who was the panegyrist of this court. The scientist and historian, al-Biruni, spent seventeen years at the court producing in Arabic an astronomical treatise, a book on mineralogy, and a great work about India. The writers were generally well rewarded and were given access to libraries taken from Ray and Isfahan.

Firdawsi, born about 932 near Tus in Khorasan, was named Abu'l Qasim Mansur, and took a pen name which means "paradise." Coming to the court of Mahmud when he was already middle aged, he continued working on his *Shah nama*, or "Book of Kings," and about 1010 completed its fifty episodes in 60,000 couplets. His epic contains a series of chronologically arranged episodes in the lives of legendary kings and heroes, progressing to those of actual historical figures. It glorifies the struggles of the Iranians against their relentless enemies of Turan, with Turan identified as the Turkic tribes beyond the Amu Darya River. Or, as he wrote, "two elements, fire and water, which rage against each other in the depth of the heart." To the Persians, it is the greatest monument of their literature, but foreigners have not been attracted to translations of the epic because of its great length and its host of unfamiliar characters. The *Shah nama* was copied and recopied in all later periods, and many copies were illuminated with miniatures by leading painters. It is written in a vigorous, direct Persian which contains relatively few Arabic words, and the language has changed so little since its composition nine hundred years ago that it can be read with ease by the people of present-day Iran. For centuries it has been recited, and even today thousands of uneducated people can repeat long passages of the poem.

The Seljuq period was brightened by the activity of a great number of poets, philosophers, and learned men. Foremost among them was al-Ghazali (Abu Hamid Muhammad ibn al-Ghazali), who died in 1111 and wrote extensively on ascetic and mystic theology. He introduced into Sufi mysticism Christian sentiment based on the concept of all-embracing love and, at the same time, brought Sufism within the bounds of orthodox Muslim belief, thus exerting enormous influence upon all later Persian thought.

Farid al-Din 'Attar lived from about 1136 until 1230. A mystic poet and a prolific writer, his major work, the *Mantiq al-tair*, tells of the search of the birds for a king.

Nizam al-Din Ilyas ibn Yusuf, born in 1140, employed the *takhallus* of Nizami. His voluminous writing is best exemplified by the *Khamsa*, or "Quintet," a collection of short epics, or idylls: *Khusraw u Shirin; Laila u Majnun; Haft Paikar*, "Seven Effigies"; *Iskandar nama*, "Book of Alexander"; *Makhzan al-asrar*, "Treasury of Mysteries."

Nizami Aruzi composed the *Chahar maqala*, "Four Discourses," about 1156: these discourses cover the occupations of administration, poetry, astrology, and medicine. 'Umar Khayyam, who died about 1123, won renown in his lifetime as a mathematician and astronomer. In the Western world he is the best known Persian poet, an eminence resulting from the high popularity of the translation by Edward Fitzgerald. This magnificent translation of many of the quatrains is not a literal one, but it is most faithful to the very spirit of Khayyam whose outlook on life was in turn pessimistic, hopeful, fatalistic, and hedonistic. While as many as 500 quatrains have been assigned to Khayyam, the majority are of doubtful authenticity. These appeared as quotations in other writings and in anthologies, none earlier than a century after his death.

Nasir-i-Khusraw was a convert to the Isma'ili sect who spent most of his life composing philosophical odes. In prose he wrote the *Safar nama*, or "Travel Diary," a most interesting account of a trip from Iran to Egypt.

The *Siyasat nama*, "Treatise on Government," by Nizam al-Mulk, mentioned earlier, concentrates on the responsibilities of kingship, illustrated with didactic historical examples and anecdotes.

The Il Khanid period saw little diminution in literary production, once the ravages of the invasions were over. Rumi (Jalal al-Din Muhammad ibn Muhammad) was born in 1207 when the Seljuq kingdom was in disarray. He travelled from eastern Iran to Qonia in Turkey where he became the head of a Sufi order, the Maulavi, renowned for its "whirling dervishes." His *Mathnavi*, in six books, is a work of the first rank, and the basic source for later Persian mystics: it stresses moral and mystical verities.

Sa'di, a native of Shiraz, brought to that town a renown that would be rivaled only by the works of Hafiz, another Shirazi. He is believed to have lived between 1184 and 1292. About 1226 he left Shiraz on travels that took him to Iraq, Syria, North Africa, Arabia, Central Asia, and India; his writings reflect these journeys, including the fact that he was for a time a prisoner of the Crusaders in Tripoli. About 1250 he returned home and in 1257 completed the *Bustan,* or "Orchard," in verse, and in the following year the *Gulistan,* or "Rose Garden," in prose with poetical passages. Critics assert that it was Sa'di who raised lyric poetry to a height which challenged that of the more formal ode, and provided an example for the writings of Hafiz. His works are basically didactic, overflowing with histories, anecdotes, fables, and dissertations in poetry and rhymed prose. Himself a Sufi, he preached moderation and advocated a virtuous life, and as ethical instruction his works retained great popularity over the centuries.

Rashid al-Din Fazl Allah, vazir of the Il Khan rulers, has been mentioned earlier. The major work, the *Jami' al-Tawarikh,* or "Compendium of Chronicles," was in four parts: the so-called *Tarikh-i-Ghazani,* a history of the Mongols through the reign of Ghazan Khan; a life of Oljeitu, which has not survived; a genealogy of prophets, caliphs, and kings; and a geographical compendium, which has not survived. Many copies of the work were made at Tabriz, and illustrated by artists in residence there.

Several major figures were active during the Timurid period. Shams al-Din Muhammad Shirazi, whose *takhallus* was Hafiz, the "Rememberer," a term applied to those who have learned the Qoran by heart, was born in Shiraz in 1326. Quite in contrast to Sa'di, he was such a homebody that he left it only once, and then briefly, before his death in 1390. He enjoyed the favor of the local court, witnessed its destruction, and was present when Timur entered Shiraz in 1388. Hafiz is reported to have been summoned to the conqueror's presence, and two accounts of what transpired have been preserved. According to the first, Timur reproached the poet for these lines:

Agar an Turk-i-Shirazi ba-dast arad dil-i-mara
be khal-il-Hinduyish bakhsham Samarqand u Bukhara-ra

If that unkindly Shiraz Turk would take my heart within
her hand,
I'd give Bukhara for the mole upon her cheek, or
Samarqand.

It was Timur's contention that he had conquered many lands in order to embellish those cities which Hafiz would give away for a trifle. The poet replied that it was through such prodigality that he had fallen into poverty. According to the other account, Hafiz said that one line had been misquoted, and that it actually went:

ba khal-i-Hinduyish bakhsham seh man qand u do
khormara
I'd give for the mole upon her cheek three mans
[weights] of sugar and two of dates.

His *Divan* numbers nearly 700 poems, nearly all of them ghazals. To the Persians he is their greatest lyric poet and is praised as the "Tongue of the Unseen," and the "Interpreter of Mysteries," because of the Sufism which pervades these lyrics. The tomb of Hafiz in Shiraz, set within a flowering garden so like those he described, is the place where people from all Iran come to take a *fal*, or "augury," from his *Divan*. With the text opened at random, the verse on which the reader's eye first alights answers the question in his mind.

The last great figure of Persian literature was Jami (Mulla Nur al-Din 'Abd al-Rahman ibn Ahmad Jami). Born in 1414, he came to the court at Herat where he enjoyed the favor of Sultan Husayn Baiqara. Possessed of a thorough grounding in theology, grammar, prosody, and music, he produced 45 separate works. In addition to many, many odes and lyrics, he composed seven epic poems, known collectively as the *Haft Aurang*, or "Seven Thrones." Also, he wrote works on Arabic grammar, composition of poetry and prose, music, the lives of the Sufi saints, and exegesis of the Qoran. During his life he was held in the highest esteem by his contemporaries and by rulers of distant kingdoms. No other writer of the period was preeminent in as many fields or so vividly expressed the mystical and pantheistic thought of Iran.

Of the other literary figures of the period only a few will be mentioned here. 'Ali Shir Nawa'i was the patron of such men as Jami and Bihzad and became equally famous as musician, painter, and poet. Nizam al-Din Shami wrote a *Zafar*

Nama, or "Book of Victory," which was the only account of Timur's conquests to be completed while Timur still lived. For its material the author, who had been involved in fighting at Baghdad and Aleppo, was given access to official records and documents. The much better known *Zafar Nama* of Sharaf al-Din 'Ali Yazdi, completed in 1424, is swollen with exaggerated praise of Timur, but is based largely upon the preceding work. Mirkhwand was the literary name of the author of the *Rawdat es-Safa*, a history of Iran from the period of the pre-Muslim kings down to 1506.

There was a dearth of notable poetry during the Safavid period. Many philosophical treatises were written by Mir Abul Qasim Findariski, Shaykh Baha'i, and Mulla Sadra and his numerous disciples, but the work had little originality and tended more toward a detailed reworking of Avicenna's earlier interpretations of the teaching of Aristotle. A severe clerical control forced all written thought to conform to orthodox doctrine. Sufism had finally lost its driving force and became repetitious and sterile. A number of mystical poets reiterated their belief that there was no real value in the world, in action, in courage and perseverance, and urged a placid submission to destiny and escape from the disillusionments and misfortunes of an active life into a personal world of fantasy.

In poetry and prose content became subordinated to form of expression. The chief delight lay in the use of ornate and elaborate language, and figures of speech were highly developed. A few of the devices used in this and later writing included homonyms, anagrams, palindromes (verses which could be read either forwards or backwards), adornment (the arrangement of verses in geometrical shapes), quadrilaterals (the arrangement of verses in a rectangle so that they could be read either horizontally or vertically), suppression (the deliberate avoidance of the use of a given letter of the alphabet), and enigmas (in which numerical dates are obtained from the sum total of the assigned numerical value of the letters in certain words).

During the nineteenth century Habib Allah Shirazi, with

[93]

the *takhallus* of Qaʻani, was very highly regarded and remains popular. He wrote many *qasidas* and also the *Kitab-i-Parishan*, a work in the manner of Saʻdi's Gulistan.

A special category of literature includes the books of ethics, *akhlaq*, and books of counsel, *pand nama*. The earliest of these writings were direct descendants of productions of the Sasanian period. Earliest in date to have survived is the *Marzuban nama*, believed to have been composed in the first half of the eleventh century by Marzuban ibn Rustam, a ruler of an area along the Caspian Sea. Originally written in a local dialect, it was turned into Persian a century later. A prince seeking to persuade his brother, the king, that he is not plotting his overthrow presents his argument with fables and anecdotes in which animals, birds, fish, and humans have active roles. In the same region the king of another local dynasty, ʻUnsur al-Maʻali Kay Kaʻus ibn Iskandar, ruled from 1049 until 1082 and crowned his career with the *Qabus nama*, a book of counsel addressed to his son. He pointed out that the fashion of the time was for a son to ignore his father's advice because he believes that his knowledge surpasses that of his elders. He wrote that a ruler should practice moderation in wine drinking, playing backgammon and chess, making love, hunting, and polo playing.

Not long before his death in 1111 al-Ghazali composed his *Nasihat al-Muluk*, "Book of Counsel for Kings," which echoes portions of the *Qabus nama* and the *Siyasat nama*. As did other writers who offered advice to rulers, he justified royal absolutism on the grounds of the need of the state for security.

Nasr al-Din Tusi, after service with the Ismaʻilis, entered that of Hulagu, and was charged with the erection of an astronomical observatory. A man of vast learning, he had previously completed his *Akhlaq-i-Nasiri*, the "Ethics of Nasir." He wrote that in justice lies the order of the realm and that, after justice, no virtue is greater than that of beneficence.

In Shiraz a contemporary of Hafiz, by name Nizam al-Din ʻUbaid-i-Zakani, composed his *Akhlaq al-Ashraf*, the "Ethics of the Aristocracy," a bitter parody of the usual manuals of ethics. It derided the accepted virtues, insisting that they had become outmoded. For example, formerly the throne was maintained by stern justice and punishment, but today justice is too merciful, and, as a result, subjects do not obey their kings, sons their parents, or slaves their masters.

Translations into English of some of the literary works mentioned in this section, as well as general histories of Persian literature in English, are included in the bibliography.

Architecture and the Arts

Until very recently the architecture of the Median period was known only from literary and pictorial sources. Fascinating descriptions of the Median capital of Ecbatana and of its fabulous palaces may be found in the works of the Greek historians, Herodotus and Polybius. There are clear-cut pictures of towns and of buildings incised on the stone reliefs which were originally in the palace of Sargon at Khorsabad. The surviving written accounts of the eighth campaign of King Sargon of Assyria against northwestern Iran in 714 B.C. aid in the understanding of the reliefs. Villages protected by ditches and lofty walls of stone are shown. The temple of Khaldia at the village of Musasir is pictured with a façade of six columns and a gable roof. Another town displays structures of squared masonry in alternate courses of black and white stones.

Now, however, at the site called Nush-i-Jan, to the southeast of Hamadan, excavators have uncovered a complex of quite well preserved Median structures, all built of mud bricks. Probably dating from the middle of the eighth century B.C., the structures include a fire temple with its fire altar still intact, a fortress with a number of long, narrow rooms, and a rectangular hall with twelve columns.

The outstanding sites of the Achaemenid period are Persepolis and Pasargadae, to employ the names used by the Greeks. It was at Pasargadae, probably called Pasragada by the Persians, that Cyrus the Great established his capital about 546 B.C. Its structures are scattered over a well-watered plain and the area may have been surrounded by a wall, to judge from the remains of a single gatehouse. Conspicuous is the tomb of Cyrus, a monumental replica in stone of a rectangular gable-roofed house, set on a stepped stone platform. According to Greek writers, the tomb bore this inscription: "O man, I am Cyrus who founded the empire of the Persians and was king of Asia. Grudge me therefore this monument." Excavations have brought to light the remaining walls and column bases of structures known as the audience palace, the residential palace, and the garden pavilion. The palaces were rectangular in plan with central columnar halls. Nearby is the enigmatic stone tower called the Zendan-i-Sulayman, "Prison of Sulayman," the prototype of the later Ka'ba-i-Zardusht. Other elements include a terraced citadel and a so-called sacred precinct. Plan forms and methods of construction represent an experimental phase preceding the more integrated architecture of Persepolis.

Persepolis, more correctly Parsa according to one of its inscriptions, rises above a wide fertile plain on a vast rectangular terrace with massive retaining walls of limestone blocks. Construction was begun by Darius about 520 B.C. and continued for nearly a century under Xerxes I and Artaxerxes I, and much later under Artaxerxes III.

The structures on the terrace fall into four general groups: gateways, audience halls, private quarters, and storehouses. Tablets from the site indicate that at one period more than a thousand workmen were busy. Artisans were brought from Babylon, Egypt, Syria, and Asia Minor and materials of construction and decoration from even more of the lands of the empire. Specific features which reflect the arts of the homelands of the artisans were well integrated into a composite style. Columns, doorways, and reliefs were of limestone, while the walls were of mud brick. Flat roofs covered all the structures.

The terrace is attained by a monumental, double reversing stairway in two flights. Just ahead is the gateway "All Lands" of Xerxes, a square chamber with four columns, flanked on two sides by colossal figures of winged bulls. To its south is the *apadana*, or "audience hall," built by Darius and Xerxes. Its great hall is nearly 200 feet square with six rows of six columns—thirteen towering columns still stand. On its north and east façades are staircases. Their almost identical reliefs depict the No Ruz ceremony. Dignitaries and Persian,

Median, and Susian guards appear on one side of each stair-case, and on the other are tributary groups from twenty-three lands of the empire, each group in distinctive garb and presenting the animals and products of their countries. In the center of each façade stood a relief depicting Darius on his throne, faced by dignitaries and backed by his son Xerxes.

From the gateway "All Lands" a processional way led east to an unfinished gateway of Artaxerxes I which gave access to a structure known as the Throne Hall because of its reliefs; begun by Xerxes, it was completed by Artaxerxes I. It was somewhat larger than the *apadana* and had 100 columns. Between the Throne Hall and the *apadana* stands the so-called Triple Portal which gives access to the southern area of the terrace. Beyond is the *tacara*, or palace, of Darius, and farther to the south the less well preserved *hadish*, another term for palace, of Xerxes. Most of the rest of the terrace is taken up by the so-called Harem, the so-called Treasury, and storerooms. Presumably all of this area contained the treasures of the realm. The burning of Persepolis by Alexander, whose soldiers had previously looted its treasures, actually helped to preserve its remains from other looters—those of stones and columns—of later ages.

Seven rock-cut tombs of the Achaemenid rulers survive. That of Artaxerxes II is in the hillside above Persepolis, and those of Artaxerxes III and Darius III are to the south of the site. A few miles to the northwest of Persepolis, at Naqsh-i-Rustam, are those ascribed to Darius, Xerxes, Artaxerxes I, and Darius II, although only that of Darius names the ruler in its inscriptions. Their façades have a row of engaged columns supporting an entablature. The composition is that of a wooden house façade transferred into stone. In front of the tombs at Naqsh-i-Rustam stands the so-called Ka'ba-i-Zardusht, or "Shrine of Zarathushtra." A stone tower with blind windows and slot-like decorations, it is uncertain as to whether it was a tomb, a fire temple, or the repository for records.

Excavations at the mound of Susa have laid bare the plan of a palace complex begun by Darius and continued by his successors. With large interior courts and long narrow rooms, certain sections of the palace were very like Assyro-Babylonian prototypes. The complex includes an *apadana* of Darius, and, from the time of Artaxerxes II a magnificent frieze of archers executed in bricks glazed with a number of colors.

The composite and eclectic character of Achaemenid art is well reflected in carvings in stone of animals, human heads cast in bronze, statuettes in gold and silver of men clad in Persian dress, and small animal figures in bronze and precious metals. Animal forms in characteristic poses were cleverly used as handles for drinking vessels and bowls, giving evidence of both vigor and imagination. There are also gold and silver bowls ornamented with rosettes and lotus buds and flowers.

The jewelry of the period—armlets, bracelets, necklaces, and earrings—has come down to us in considerable quantity. It is best represented by the famous Oxus Treasury, a collection of gold and silver jewelry set with pearls, lapis lazuli, and colored stones, most of it now in the British Museum.

Achaemenid cylinder seals and seal stones, many bearing scenes of religious worship or wild animal hunts, have survived in large numbers. There are also gold and silver coins of imperial issue ranging in date from Darius I to Darius II, each bearing a portrait of the king, who is always armed with a bow and carries either a spear, a dagger, or arrows.

The excavations of the Oriental Institute of the University of Chicago and later of the Iranian government at Persepolis brought to light in the debris of the royal treasury many interesting objects, notable among which were the stone and alabaster plates and bowls brought back from Egypt by the victorious Achaemenid army.

Since the major Seleucid cities lay to the west of the Iranian plateau, few monuments exist in Iran. Kangavar, northeast of Kermanshah on the ancient highway, has the remains of a once immense temple believed to have been dedicated to Anahita, while the ruins of a much smaller temple are found at Khurha, near modern Qum.

Nearly all the architectural remains of the Parthian period are also to the west of Iran, and reflect Hellenistic and Mesopotamian influences tempered by traditions of local craftsmanship. Monuments above ground or excavated ruins survive at

Dura-Europos on the Euphrates, at Hatra, Seleucia, Assur, Nippur and Warqa in Mesopotamia, and at Nisa and Kuh-i-Khwaja within Iran. These include palaces and houses, religious structures, and tombs, constructed of baked brick and stone. The only element of these structures that will be described is the *ivan*, the element which became an important feature of the later architecture of Iran—a long rectangular tunnel-vaulted hall with open façade and rear end closed by a wall. The Parthian ivans and other vaulted structures represent a transition from the post-and-lintel construction of the Achaemenid and Seleucid periods to the vaulted architecture of succeeding periods.

Parthian art was even more eclectic than the architecture and, indeed, the term Parthian art has little precise meaning. Objects found in different parts of the empire often have little in common. For example, in the Tehran museum are several small human heads carved in stone, one of them purely Hellenistic in style, one definitely Buddhistic, and the others Graeco-Bactrian. The museum also houses a magnificent life-size statue in bronze of a chieftain or ruler: the fine head with its long hair and heavy moustache was cast separately from the body. Probably, as was the case in the Achaemenid period, skilled artists were brought to the major cities of Parthia from every corner of the empire. In general, the art of the period uses Hellenistic forms regardless of the content originally associated with them, never managing to assimilate western influences and to recombine them in a fresh and original way.

Stucco, a material which was later to be highly developed in Iran, began to be used on buildings in friezes with incised geometrical and floral patterns. Terracotta, used for coffins and religious figurines, was even more popular, while pottery vessels coated with a green glaze anticipated later work in this field. Ornamental stone carving on architecture employed direct copies of Hellenistic motifs. Wall paintings of the period have survived: those at Dura-Europos have been studied in detail as have the paintings from Kuh-i-Khwaja in Sistan, which are Graeco-Bactrian in style, and apparently date from the first century A.D.

The architecture of the Sasanian period is represented by palaces and residences, fire temples, fortresses, dams and bridges, city plans, and special memorials. Quite a number of these buildings stand in fair condition, but the problem of dating them chronologically within the period is not yet solved. Most construction was of stone masonry, but baked brick was employed in domes and, less frequently, in lower walls.

At Firuzabad, Shapur, Bishapur, and Sarvistan in Fars, at Ctesiphon on the lower Tigris, and at Qasr-i-Shirin on the present frontier between Iran and Iraq are the remains of great palace complexes. The Firuzabad palace, built in the time of Ardashir, is a rectangular building over 180 feet wide and more than 300 feet in length. In the center of the façade is a great ivan which leads into a square-domed throne chamber, and back of this chamber is an interior court surrounded by living quarters. All the rooms are covered either by domes or tunnel vaults. At Sarvistan, some sixty miles southeast of Shiraz, is a well-preserved palace which was probably built in the fifth century. The plan differs from that of Firuzabad: there are fewer and larger rooms and a number of entrances. Neighboring ruins suggest that the main palace was surrounded by gardens and other buildings. The ruins of an extensive city at Bishapur include a palace of Shapur I. Its main hall, cruciform in plan, consisted of a vast square chamber, originally surmounted by a dome, with ivans in the corners of the cross.

At Qasr-i-Shirin are the ruins of a huge complex covering an area an eighth of a mile wide and a quarter of a mile long, the main palace structure raised above the ground on a maze of tunnel-vaulted rooms. This work dates from the very end of the Sasanian period.

At Ctesiphon the salient feature of the palace area is the great vaulted ivan of the Taq-i-Kisra, so named by Arab historians, which rises to a height of 90 feet and is 75 feet wide and 150 feet deep. On its flanking façade walls is a poorly organized system of superimposed arcades which copy Hellenistic façade compositions. Later the Arabs tried in vain to pull it down, and in Islamic times builders tried to erect structures large enough to overshadow it. During the reign of Khusraw the ruler held audience in this hall, seated on a throne behind a hanging curtain. On special occasions the curtain was drawn to reveal him on a splendid throne with the massive regal crown suspended over his head on a chain hung from the vault above.

The primary role of the sacred fire in the religion of the Sasanians has already been mentioned. In recent years over a score of well preserved fire temples have been recorded. Apparently, in many of them the sacred fire burned within a temple, while those which appear to crown rocky heights burned in the open air and were visible from a long distance. The basic element of the fire temples is a square dome-crowned chamber which, whether an isolated unit or surrounded by a corridor, normally had a wide arched portal piercing each wall, the term for this plan type being carried over into the Islamic period as the *chahar taq* or "four arches." The dome over the chamber sprang from four arches or squinches which bridged the corner angles. It is noteworthy that the world of Rome and Byzantium developed the form of the pendentive or spherical triangle to support a dome over a square plan, but the eastern and Islamic world has continued to use the squinch form down to the present day. The structural materials of the period were stone and fired brick and wall surfaces were frequently coated with plaster, as in palaces excavated at Damghan where incised plaster decoration was used extensively.

Architecture and the fine arts clearly reflect a strong reaction against earlier Hellenistic influences and even give evidence of a studied effort to re-use traditional native forms. Little conscious development of forms and motifs took place, while the reliefs seem to reveal a stylistic conflict between those which display a static monumentality in the figures and those in which a forceful realism is present.

More than twenty very large reliefs were carved on vertical rock faces. Most are situated in Fars, either above a spring-fed pool or along the bank of a river, although six occur below the tombs of the Achaemenid rulers at Naqsh-i-Rustam. The scenes, a number crowded with many figures, depict the investiture of a king by Ahuramazda or Anahita, the king enthroned, a ruler victorious over Parthian, Roman, and Indian enemies, and a ruler fighting lions. At Taq-i-Bustan, near Kermanshah, Khusraw II is shown on the rear wall of a grotto: he sits astride his favorite charger, *Shabdiz*, "Night Hued," clad in mail and brandishing a lance.

Very well known are the splendid silver dishes and bowls, many of which are now in Russia in the Hermitage Museum. Some show the king seated on his throne holding royal audience or enjoying the pleasures of the hunt, others show battle scenes, the investiture of the monarch, deities, vigorous animals, often in combat, and graceful birds. Vessels and ewers of bronze have also survived in quantity: notable among these are incense burners in the form of ducks or other birds. Many splendid Sasanian textiles are now found in museums and private collections, but it is not yet clear which of the many pieces assigned to this group were woven in Iran and which in the countries along the eastern shores of the Mediterranean. The pottery of the period is not particularly fine; a common type has patterns in low relief covered with a color glaze. Plaster decoration was extensively used on walls and vaults, usually in the form of repeating motifs with floral and animal forms.

In the period from 641 until 1000 Iran was working out her own particular version of Islamic architecture. One of the fundamental precepts of Muhammad's religion was the necessity of prayer as stressed in the Qoran, and before long the true Muslim believer was required to recite five ritual prayers each day. The prayers could be said anywhere, provided that the believer's face was turned towards Mecca. Group prayers were recited at noon on Friday. The need for a special place for prayer appeared very early, and it is believed that the house of the Prophet Muhammad was used as the first *masjid* or "place of prostration," the mosque. As the Arab conquests spread into far countries it became necessary to provide mosques for the soldiers. At first the mosque may have been any open space fenced in with a reed paling which could be transported when the army moved. A little later the area was determined by four arrows shot from a central point and a ditch dug around the space thus defined. Still later, covered colonnades were erected around the open space or court. In each country the Arabs also took over existing structures for use as mosques: in Syria and Palestine churches were so used and at Ctesiphon the Taq-i-Kisra was converted into a mosque. When the Muslims began to build their own permanent mosque structures they adapted the architecture in vogue in each area and made use of the local craftsmen. The essential elements of the mosque were an open court with arcades on one or more sides and a covered sanctuary area within which was the *mihrab* or "prayer niche," so placed that worshippers would always face toward Mecca.

Within Iran a wealth of monuments of the Sasanian and earlier periods offered plan forms and constructional methods readily adaptable to the new religious architecture, and from documentary sources it is known that thousands of fine buildings were erected in the early Islamic period. Of these only a handful, and not the more renowned, have survived to the present day. Many earlier structures were converted into mosques; one precise account in an early history refers to an Achaemenid palace as having been so altered.

Of the extant monuments erected in Iran prior to the year 1000, the Tarik Khana at Damghan, a mosque which was probably built in the ninth century, consists of a square court surrounded by arcades which are several bays deeper on the side which sheltered the mihrab. The *Masjid-i-Jami'* or "Congregational Mosque," at Nayriz, which may date from the tenth century, was built as a large ivan with the mihrab set into the rear wall of the hall. The tenth century Masjid-i-Jami' at Nayin is similar in plan to the mosque at Damghan but has the added feature of walls, columns, and mihrab surfaces, decorated with plaster ornamentation in Arabic inscriptions and deeply cut geometric and floral patterns.

Structures other than mosques were also built. In this early period, as well as in later centuries, some were of mud brick and others of fired brick. Naturally enough, most of the former have vanished: they included many palaces, for successive rulers had new ones erected rather than occupying existing ones.

The Ziyarid ruler Qabus erected a tomb at the southeast

corner of the Caspian Sea which is the earliest monument
with a dated inscription surviving in Islamic Iran. Written in
the blocky early Arabic script called *kufi*, it reads: "In the
name of Allah, the merciful, the compassionate; this castle
was built by the Amir Shams al-Ma'ali the Amir, son of the
Amir, Qabus son of Vashmagir, who ordered it built during
his lifetime in the lunar year 397 and the solar year 375"
(1007 A.D.). The tomb, built of fired brick, is an enormous
cylinder capped by a conical roof. The circular plan, broken
by ten flanges, is 56 feet in diameter and the walls are 17
feet thick. The height from the ground to the tip of the cone
is 160 feet. According to the historian Jannabi, the body of
Qabus was enclosed in a glass coffin which was suspended by
chains from the interior dome of the tower.

Other tomb towers still stand at Resget and Lajim along
the Caspian coast and at Damghan: the one at Resget has an
inscription in Arabic except for a last line in Pahlavi. Circular
or round in plan, and with a domed or tent roof, these tomb
towers established a type which was to persist for several
centuries.

A few monuments of the Ghaznavid period have survived.
At Ghazni the mausoleum of Mahmud has vanished but the
tombstone remains; a triangular prism of marble with beautiful inscriptions in Arabic, including the exact date, day, and
hour of his death. Also at Ghazni are two tall minarets en-
riched with decorative brickwork: one was erected by Mas'ud
II and the other by Bahram Shah. Some 300 miles southwest
of Ghazni, at Lashkari Bazaar, is a military camp with three
fortified palaces. The southern palace displays mural paintings of some fifty members of the ruler's Turkish guard. Still
further to the west, at Sangbast in Iran, is the domed tomb
chamber attributed to Arslan Jadhib, a vazir of Mahmud.

In this early Islamic period began the remarkable skill and
artistry of the pottery craft which was to make Persian ceramics the equal of the production of any other country. Nishapur, in Khorasan, was a most important city, and over a number of years an expedition from the Metropolitan Museum
of Art brought to light artifacts from the eighth through the
tenth centuries. Pottery of several distinctive types was found.

Most were rather flat bowls with colorless glazes over bold designs in several colors. Bowls with elegant kufi inscriptions around the rim were probably imported from Samarqand, the center for the production of the so-called Transoxiana wares. Other types included the forerunners of luster painted pottery, and dishes with splash glazes in imitation of Chinese wares. Nishapur also yielded a fresco painting of a prince hunting with a falcon, a subject also found on this pottery.

The silver dishes of this period are in shape, technique, and subject matter direct successors of the Sasanian silver. A few fragments of fine textiles have survived. Wood carving is represented by the double doors of mosque and tomb, by Qoran reading stands, and by examples of the *minbar*, the pulpit which stands next to the mihrab in the mosque.

A great deal of building activity was carried on during the Seljuq period and sections of several Iranian mosques date from these years. The mosques now reflect the standard plan of a central open court surrounded by arcades with an ivan on each side of the court and columned and vaulted prayer halls on both sides of each ivan. The small mosque at Zavara was constructed to this plan in a single building operation about 1135, but at most of the mosque sites only the traditional square chamber crowned by a dome was erected in this period and the ivans and arcades added later, to round out the composition. The square dome-chambers were also used for shrines and tombs: typical examples of the form are in the Masjid-i-Jami' and in the Haydariya *madrasa* at Qazvin, in the mosque at Gulpaygan erected in the reign of Sultan Muhammad, and at Barsian.

The largest, most complex, and most important monument of the Islamic centuries is the Masjid-i-Jami' at Isfahan. Two splendid dome chambers are of the Seljuq period: one was erected in 1088 by the order of Taj al-Din, a political rival of Nizam al-Mulk for the favor of Malik Shah, and the other, which houses the mihrab, about the same time by Nizam al-Mulk himself. It was long believed that these structures were a part of a mosque first erected in Seljuq times and then much added to in later periods. However, studies and excavations begun in 1970 indicate that a Sasanian structure occupied part of the site, and that in the eighth or ninth century a mosque came into being which was basically a rectangular walled court with rows of columns along each wall.

Sanjar was buried at Merv, his favorite place of residence, in a tomb constructed during his lifetime which was originally connected with a now vanished mosque. The structure is a great square 77 feet on each exterior side with walls 19 feet thick. The four walls are crowned by galleries, and behind and above the galleries rises the bold profile of the great dome. Interior wall surfaces are covered with plain plaster, and the gallery vaults with decorated plaster on which incised lines imitate brick bonding patterns and trace delicate inscriptions against a floral background.

Tomb towers of the Seljuq period also survive. Especially noteworthy are the multiple-flanged tomb of Toghril Beg at Ray, three towers at Maragha, and two at Nakhichevan. The Gunbad-i-Surkh at Maragha, dated 1147, has on its exterior walls double panels executed in a variety of bonding patterns in red brick. Other types of architectural decoration common to the period are strips of baked terracotta interlaced in geometrical patterns, and magnificent carved plaster mihrabs and inscription friezes.

The field of ceramics displayed an amazing development which was to culminate in the sometimes over-elaborate pottery of the succeeding Mongol period. Near the end of the Seljuq reign came dated luster pottery and architectural tiles as well as polychrome pottery from the workshops at Ray, Sultanabad, Saveh, and Kashan. Such centers as Aghkand, Yastkand, and Amul turned out colorful glazed pieces on which the figures of animals and birds were incised or raised in low relief by cutting away the background areas. At Ray and Kashan subjects common to later miniature painting were used and many pieces show scenes from the *Shah nama*, horsemen and the hunt, nobility enthroned, or congenial groups in flowering gardens. A number of important centers turned out vast quantities of fine cotton, wool, and silk textiles; fragments of silks show a consummate technical skill in the execution of inscription bands and human and animal figures. Metalwork of the period includes bronze mirrors whose backs are adorned with scenes in low relief. Many brass or bronze kettles, incense burners, jugs, candlesticks, trays, lamps, ewers, and pen boxes, and a smaller number of silver vessels and trays, are gathered in museums and private collections. Craftsmanship of the highest order appears in the engraved designs, which are often inlaid with silver or with both silver and gold.

Building activity in Iran seems to have fallen off at the end of the twelfth century and scarcely any monuments of the early thirteenth century have survived to the present day. The Mongol invasions destroyed countless structures in the ruined towns and it was not until thirty years after their first appearance in Iran that monumental architectural construction began to revive. Hulagu built the Maragha observatory, a treasure house, his mausoleum, and other structures, but of them all only the foundation walls of the observatory remain. However, at least a hundred important structures built from the time of Hulagu to the death of Abu Saʿid still stand in Iran. Most of these are mosques, shrines, or tombs, but there are a few caravanserais and structures of other types.

The architectural style of these buildings is a direct continuation of that of the Seljuq monuments erected nearly a century before. The use of elaborate brick bonding as a decorative feature tended to die out but ornamental plaster reached a high development, and bold, exuberant patterns covered mihrabs, inscription bands, and the surfaces of walls and vaults. Much of the plaster was tinted in red, blue, white, green, and yellow. A new decorative technique appeared with the use of glazed tiles on both exteriors and interiors. At first small pieces of light and dark blue glazed tile were used, and then white and black were introduced until whole surfaces were covered with patterns made up of small, carefully fitted pieces of glazed tile. This technique, known as mosaic faience, was to become the hallmark of the architecture of the Timurid and Safavid periods.

In 1297 Ghazan ordered construction work begun in a suburb of Tabriz and in a few years his twelve-sided tomb structure, crowned by a great dome, was the center of a group of buildings set within gardens. The buildings included a monastery, a hospital, religious schools, an observatory, a library, a palace or administration building, and an academy of philosophy. Today the site is marked by the mounds of masonry of his ruined tomb.

Outside Tabriz Rashid al-Din established a suburb named "Quarter of Rashid," devoted to the promotion of the arts and sciences, and soon theologians, jurists, traditionalists, reciters of the Qoran, students, and craftsmen of every trade were lodged in 30,000 charming houses. Copies of the universal history were embellished with pictures, and contemporary records state that the manuscript painters were held in high favor. Although some of this work continued to reflect the formulae of the thirteenth century Baghdad school of painting and unusually fine large sheets of paper were still brought from Baghdad, the style was in close harmony with the spirit and character of the court. In the miniatures the Il Khan rulers and their wives wear Chinese headdress, and a strong Chinese influence is reflected in trees and flowers dashed in with the rapid strokes common to a brush technique and clouds given the twisted and voluted eastern forms.

Early in the fourteenth century the vazir Taj al-Din 'Ali Shah erected a vast complex which included a large open court, a great pool with a central fountain, a monastery, a religious school, and the central element of the mosque. This latter feature was an enormous ivan hall a hundred feet in interior width. As first erected, it must have closely resembled the Taq-i-Kisra at Ctesiphon. Today a massive section of the thirty-foot thick walls remain.

In 1306 Oljeitu ordered work begun on the city of Sultaniya, located on a wide plain fairly near Qazvin, which was to replace Tabriz as the capital of Il Khanid Iran. The ruler himself, Rashid al-Din, Taj al-Din 'Ali Shah, and many courtiers vied with each other in financing the erection of palaces, public buildings, and entire quarters of the new city. Within its stone walls construction work was started on the mausoleum of Oljeitu which still stands in quite good condition: it is one of the finest monuments ever erected in Iran and would be a credit to any country or any style of architecture. The mausoleum is octagonal in plan. Its exterior walls are crowned by eight minarets forming a circle around the huge dome, which is sheathed with bright blue glazed bricks. The interior is a great octagonal hall 80 feet in diameter and about 170 feet high from the ground to the apex of the dome, with a small mortuary chapel on the south side. While construction work was being pushed forward Oljeitu conceived the idea of bringing the bodies of ʿAli and Husayn, the Shiʿa saints, from Iraq and turning his tomb into a shrine for their remains, and the decoration was carried out so that the name of ʿAli figured prominently in the inscriptions. Later on the plan was discarded, and the entire interior of the tomb was redecorated before the monument was completed and the new city dedicated in 1313. The bazaars of the new city soon became famous. In them could be found spices from India, turquoises from Khorasan, lapis lazuli and rubies from Badakhshan, pearls from the Persian Gulf, silk from the Caspian shores, indigo from Kerman, textiles from Yazd, cloth from Venice, Lombardy, Germany, and Flanders, brocades, oils, musk, Chinese rhubarb, sparrow hawks from Europe, and horses and hounds from Arabia.

Few dated ceramic pieces survived the years of devastation between 1220 and 1242 but the larger number of extant examples dated after 1242 testifies to the revival of work in ceramics at Kashan, Ray, and other centers. The dated pieces from Kashan include great mihrabs assembled from several large pieces of luster tile, luster star tiles which were used with cross tiles in lining the lower interior wall surfaces of buildings, and fine luster or polychrome painted bowls, dishes, and jugs. The luster pieces have a red, golden, or brown transparent glaze obtained by covering the finished piece of pottery with a pigment of metallic copper or silver salts and then re-firing the piece so that a thin film of metal is deposited on the surface.

In general, the pottery is more ornate and combines a greater variety of techniques than was common in the Seljuq period. Characteristic of the Il Khanid period is the so-called Minaʾi ware in which polychrome patterns were painted over the glazed pottery. Richest of all was the polychrome overglaze pottery on which parts of the design were gilded. Although certain distinctive pottery types have been assigned to the busy workshops at Kashan, Ray, and Sava, many other towns also produced fine ceramics. In recent years a great many pieces have been found in excavations along the southeastern corner of the Caspian Sea.

Some mention has already been made of the style of the illuminated manuscripts of the history of Rashid al-Din produced in his suburb outside Tabriz. Extant pages from an illustrated work of al-Biruni are dated 1307 and may have been painted at this same site. In this period also copies of the *Shah nama* were illuminated and the large pages formerly in the Demotte Collection, believed to have been painted in Tabriz. In these pages the Far Eastern influences are already less strong: compositions and details have been altered, the colors are stronger, and gold backgrounds are now in vogue. Other centers of miniature painting were probably at Shiraz and Herat. The splendid ornamented pages of a Qoran which was made for Oljeitu in Hamadan in 1313 are covered with geometrical and floral patterns very like the patterns used in contemporary architectural decoration, as are pages from Qorans written at Mosul and Maragha.

Textile weaving was carried on at Tabriz, Nishapur, Herat, Qum, Yazd, and other centers. About thirty fine silks, some of which show a definite Chinese influence, have been preserved in museums and private collections. Some make use of gilded or silvered thread, many are woven in narrow striped patterns, and still others are covered with palmettes, lotuses and other floral forms. Some of these silks, one of which bears the name of Abu Sa'id, were found in Europe far from Iran.

The name of Abu Sa'id is also found on a magnificent brass basin inlaid with gold and silver. Other notable examples of the metalworkers' art continue the style of the Seljuq period. The wood carving of the period is known to us from a number of minbars and doors of mosques and shrines in which two main types of pattern were used: geometrical patterns of interesting polygons, and intertwined conventional plant forms.

Many architectural masterpieces of the Timurid period, known to us from contemporary accounts, have vanished from the face of the earth. One such monument was the great mosque built by Timur at Samarqand, and only one section remains of the fabulous palace of Timur at Kesh which took twenty years to build. However, at Samarqand a number of monuments survive. Foremost is the Gur-i-Amir, the tomb of Timur, with its great ribbed dome rising from a high drum. The so-called Bibi Khanum mosque is actually the Masjid-i-Jami' built by Timur, while the Registan, a great open square, is fronted by a madrasa of Ulugh Beg and two later structures. The ensemble called the Shah-i-Zindah comprises a series of mausolea along a narrow street, with an entrance portal of the time of Ulugh Beg.

Gawhar Shah, wife of Shah Rukh, patronized an outstanding architect, Qavam al-Din Zayn al-Din Shirazi. Between 1405 and 1417 he erected the splendid mosque which bears her name within the shrine complex of the Imam Reza at Mashhad. Later he constructed a series of structures at Herat of which only her mausoleum, a number of minarets and some shattered walls remain. Better preserved is a madrasa at Khargird in Khorasan.

Other standing monuments of the fifteenth century include the imposing shrine of Khwaja Abu Nasr Parsa at Balkh, the shrine of Khwaja 'Abdullah Ansari outside of Herat, the shrine at Turbat-i-Shaykh Jam and the tomb of Zayn ad-din at Tayabad, the latter two just within Iran's present eastern frontier.

Most of these structures, as well as those mentioned earlier, reflect an interest in large-scale architecture which is most evident in the great height of the principal ivan of each monument. Nearly every building of the period was gorgeously clad in multi-colored faience, and near the end of the period the glazed material was used on the interior as well as on the exterior surfaces. In western Iran is a structure whose decoration links the Timurid and the Safavid periods. This is the so-called Blue Mosque at Tabriz in which Jahan Shah, the most powerful ruler of the Qara-Qoyunlu dynasty, was buried in 1467, two years after its completion. The main area of the building is rectangular in plan with a wide vaulted passageway around three sides of a large dome chamber which here took the place of the usual open court. Behind the dome chamber is a slightly smaller domed sanctuary. The exterior walls and entrance portal of the damaged structure have brilliant mosaic faience while on the interior the bright glaze is set against a background of reddish-buff brick.

Ceramic production seems to have been of minor importance in the Timurid period, but the marvelous development in book illustration is the major feature of fifteenth century art. The principal schools of painting were at Shiraz and Herat, the less important Shiraz school tending to continue the earlier traditions of painting at Baghdad and Tabriz.

At Herat the production of manuscripts was given high priority. For example, Baysunghur maintained a staff of forty which included calligraphers, painters, illuminators, gilders, and binders. The classic works of Persian literature were copied and illustrated. Herat painting reflects the influence of individual masters of conspicuous technical skill and replaces the rigid formalism of earlier periods with a studied naturalism. The work of this century represents the climax of development of Persian miniature painting. Compositions use a decorative rather than a realistic grouping of the figures. Details of architecture, costumes, and growing forms are executed precisely and elaborately. Although landscape is always treated as background for subject matter rather than as dominant motif, mountains, trees, and sky are painted with loving care. The subjects seem crystallized into static and unlifelike scenes, for the artists avoided true perspective, the use of shadows, and plastic moulding of figures, and the faces are usually void of emotion. The colors are very brilliant and pure and are applied in flat tones.

Bihzad, born about 1440 and still active under the Safavid ruler Shah Tahmasp, has always been considered the greatest Persian miniature painter and, of course, had a marked influence upon contemporary as well as later artists. Countless miniatures were once ascribed to his hand, but scholarly study has narrowed the field to a number of signed pages and a few other pictures which bear the unmistakable mark of his personal style. A copy of the *Bustan* of Sa'di, now in the Egyptian Library at Cairo, contains four miniatures each signed "work of the slave Bihzad," and a fifth double-page frontispiece representing a banquet of Sultan Husayn is certainly by the master. A Nizami manuscript also has three small pictures signed by Bihzad. A *Zafar Nama*, written in 1467 for Sultan Husayn Bayqara, contains six magnificent double-page paintings of such scenes in the life of Timur as the king enthroned in a garden, or supervising the construction of a mosque. These paintings are unsigned but are generally assigned to Bihzad.

Bihzad must have had many pupils in Herat, and some of the work of one of them, Qasim 'Ali, has survived. Other pupils and painters who drew inspiration from his work moved to Bokhara and set up flourishing workshops in that city. One of these artists, Mahmud Muzahhib, painted a fine portrait of 'Ali Shir Nawa'i which is still in existence. The school of Herat also attracted painters who were active before Bihzad reached his prime or who were not under his direct influence. Among the first group was Nasrullah Abu'l-Ma'ali, who illustrated a manuscript of the fables of Kalila and Dimna which is now in the Gulistan Palace at Tehran. There is also a copy of the *Shah nama* with splendid pictures by an unknown hand. One of the most charming miniatures ever painted in Iran is associated with the Herat school and possibly with the painter Ghiyath ad-din, who accompanied an embassy sent by Shah Rukh to the court of China and kept a diary of his travels. The painting, done on silk, shows the meeting of the Prince Humay of Iran with the Princess Humayun of China, in a garden bursting with fruit trees in blossom and a whole catalogue of exquisitely drawn flowers.

Sustained architectural activity was a characteristic of the Safavid period. The great number of structures erected in this period which still stand in good condition includes scores of small shrines in hidden villages. Plan forms, structural methods, and materials used are all continuations of the work of earlier periods. The keynote of the period was color, and polychrome mosaic faience now clothed both interiors and exteriors of the principal monuments. The task of cutting and fitting and assembling many thousands of small glazed pieces which, on a single building, might take several years, finally proved to be so slow and so costly that it was superseded by a method of comparable effect. In this technique, called *haft rangi* or "seven colors," the details of a large decorative panel were painted with as many as seven different pigments on square tiles, and the colors then fired on the tiles in a single operation. The use of the double dome also characterized the period. Square chambers were covered by a shallow interior dome above which rose a high, more bulbous dome, with the open space between filled with an intricate supporting framework of timbers.

At Isfahan Shah 'Abbas undertook the construction of a new imperial city adjacent to the ancient town. The first stage of the work established the main lines of the general plan and saw the erection of the buildings essential for the domestic, civil, and religious requirements of the court. The second stage, beginning in 1611, elaborated the existing elements and duplicated the earlier structures in far larger and finer buildings. Today these splendid monuments remain in good condition, and Isfahan, of all the cities in the country, offers the visitor a complete picture of older Iran. The Maidan-i-Shah, or "Imperial Square," is a vast rectangle. At its southern end rises the magnificent Masjid-i-Shah, the "Imperial Mosque," with its towering façade flanked by minarets, its great dome, and its dazzling inner courts, with all surfaces clad in faience. At its northern end a lofty entrance to the old covered bazaar was built. On the east side of the square is the Masjid-i-Shaykh Lutfullah, a jewel of a structure, begun in 1603 and completed in 1619: it took many years to cover the fabric of these structures with tilework. Opposite the mosque stood the 'Ali Qapu, the "Lofty Gateway," a skyscraper of six storeys which was the monumental entrance to a spacious garden area strewn with palaces and pavilions. To the west of this area ran the avenue called the Chahar Bagh, or "Four Gardens."

A particularly charming feature of the period was the setting of brightly colored buildings in pleasant gardens in reference to pools and streams. In such gardens were erected shrines of saintly men; a typical example is the tomb of Muhammad Mahruq at Nishapur, where the same garden shelters the remains of 'Umar Khayyam. Many palaces built by Shah 'Abbas and other rulers of the line were in the royal gardens. At Isfahan one may still see the palaces called Chihil Sutun, Talar Ashraf, and Hasht Bihisht, while others still stand along the Caspian coast at Amul and Ashraf. The rulers enjoyed living in the open air, as the following account testifies: "For thirtie dayes continually, the King made that Feast in a great Garden of more than two miles compasse under Tents, pitched by certaine small courses of running water, like divers Rivers, where every man that would come, was placed according to his degree, eyther under one or other Tent, provided for abundantly with Meate, Fruit and Wine, drinking as they would, some largely, some moderately without compulsion. A Royaltie and Splendor which I have not seene, nor shall see again but by the same King."

Carpet and textile weaving reached almost incredible perfection in the Safavid period. Great carpets woven in the royal workshops and destined for palaces or shrines were fabulously expensive to produce, and as we see them today hung on the walls of museums it is difficult to believe that such priceless objects were actually made to be walked upon. Nature and the flowering garden furnish the pattern themes. Some of the patterns most frequently employed were hunting scenes; entire gardens as seen from directly above; compositions of a central medallion surrounded by a field of floral scrolls, arabesques, and animals; fields of rows of flower-filled vases; compartmented flower-strewn fields; and, all-over foliage designs. The earliest preserved carpets of great distinction date from the opening of the sixteenth century, and the craft remained at its peak throughout the entire century.

Some of the most renowned include two hunting carpets, one in the Boston Museum of Fine Arts and the other in the Poldi Pezzoli Museum in Milan, and a medallion and arabesque carpet, more than 26 feet long and 13 feet wide, in the Metropolitan Museum of Art. Also, the huge medallion carpet with hanging mosque lamps formerly believed to have come from the shrine of Shaykh Safi at Ardabil and now in the Victoria and Albert Museum, London: it is signed by its maker, Maqsud of Kashan, and dated 1539.

Later in the century the workshops at Kashan wove a series of silk rugs; many small ones and three huge pieces have survived. Still more luxurious were the so-called Polonaise carpets whose patterns were interwoven with silver and gilded silver threads. Prayer carpets using the mihrab as the central theme were also woven in quantity in this period.

The textile weavers, working at Isfahan, Yazd, and less important centers, turned out silk twills, satins, velvets, silks using silver and gold thread in patterns only, and silks in which the entire background of the design is of metallic thread. Pieces from the period of Shah 'Abbas survive in quantity and are of first-rate technical skill. Ghiyath al-Din 'Ali of Yazd is the best known of the weavers who signed their work; he and his fellow workers at Yazd specializing in panels enclosing a single blooming plant set within an arched frame. A very popular theme was that of repeated floral sprays against solid grounds of salmon pink, deep blue, yellow, or green. Equally popular was the garden scene which often represented personages bearing a jug and a wine cup, cypress trees, flowering bushes, birds, animals, and a pool. Most of these stuffs were destined for costumes to be worn at the royal court, and a number of these Safavid costumes survive to the present day.

Shah 'Abbas assembled, in a building in the shrine complex at Ardabil, a very large collection of Chinese blue and white porcelain—which is now in the Tehran Museum—and much of the pottery of the Safavid period followed Chinese models, using the delicate colors and refined designs of the Far East. Shah 'Abbas even went so far as to bring three hundred Chinese potters and their families to Iran to instruct the local craftsmen. Persian porcelain was produced in quantity but never rivaled the genuine Chinese work. Celadon ware, for example, was made in imitation of the Chinese type but the best of the Persian manufacture was a gray-green not true in tone to the master models. A typical Persian ware of the period is the so-called Kubachi pottery, many pieces of which carry a rapid, sketchy drawing of the human figure with the designs painted in blue, green, red, and tan on a cream slip and covered with a transparent glaze.

12. Typical mountain landscape
north of Tehran

13. The outskirts of a typical farming village with its irrigation channel

14. One of hundreds of the sturdy old brick bridges of Iran: the bridge at Qaflan Kuh

15. Mashhad. A stalactite-filled vault in the Mosque of Gawhar Shad

Painting continued to be of a very high quality. At Qazvin Shah Tahmasp was a munificent patron, and a marvelous *Shah nama* produced for him there contains 275 very large miniatures; 75 of them are now in the Metropolitan Museum of Art. The head of the library staff was Mir Musawwir, a painter whose known output includes some of the illustrations in the Tahmasp *Shah nama*. At Isfahan under Shah 'Abbas, the leading painter was Reza-i-'Abbasi whose output also included the calligraphy for the inscriptions on some of the monuments. European influence and dress came into the art as the result of the presence of Western artists at the court.

With the downfall of the Safavid dynasty artistic production declined. At Shiraz architecture enjoyed a fine revival during the reign of Karin Khan Zand. In the Qajar period buildings reflected the influence of Russian architecture, and painting displayed European influence as large oil paintings of the rulers, painted with a realism foreign to earlier periods, came into fashion. In one field the jewelers of the Qajar period excelled: this was the making of vases, dishes, boxes, and water-pipe fittings of enameled gold and silver, decorated with floral designs, figural scenes and portraits in enamel. There was also an astonishing productivity of *qalamdans*, lacquered cases for holding reed pens and inkwells, which were decorated with a variety of painted scenes.

Social Patterns

Iran's social structure was built around the rule of an absolute monarch supported by feudal lords, who as early as the Achaemenid period were named governors in control of vast areas, and in this and later periods their power tended to become hereditary and therefore confined to relatively few families. The loyalty of the feudal lords was always a somewhat precarious quality, and at every period the monarch was faced with the task of retaining their military support while keeping them from growing too powerful.

In Islamic times, with the extension of stable government and a carefully organized system of tax collection, a large bureaucracy grew up, staffed, in part, by men of learning. Thus, men of the pen ranked high within the society, followed by the warriors, the traders and artisans, and, finally, the agriculturists. By the seventeenth century, landowners were distinguished from the working farmers, and the civil and military authorities were attached to the court. In general, the court and the religious hierarchy were above these classes, constituting the forces that guided and controlled the community of believers.

In time the warrior class, represented by the levies furnished by the feudal lords, lost out to the recruitment of tribal forces, who were, in their turn, replaced by mercenaries serving the central authority. As foreign invasions ceased after the Safavid period, internal security grew and the importance of large landowners increased. The electoral law of 1906 divided the voters into six classes: princes and members of the royal family, religious figures, nobles, bureaucrats, merchants, and farmers. The nobles included individuals whose loyalty to the throne had been rewarded by titles, grants of land, and exemption from taxes: this group grew in size during the nineteenth century.

A persistent division within the Persian community was between the tribal nomads and the settled people. Both elements retained the ancient structure of family, clan, and tribe, but among the settled it tended to decay as people changed their habitats and severed old ties. In sociological terms the Persian family is described as being extended, patrilineal, patrilocal, patriarchal, endogamous, and occasionally polygamous. The patriarch shared with other males the responsibility for the welfare of the family, and the primary tie of the individual in this society was to his family.

More hardy and warlike than the settled Persians, and often of a different ethnic stock, the tribes won a part of their livelihood from raiding the trade routes and villages, even attacking and capturing towns. Such dynasties as the Seljuq, Afshar, Zand, and Qajar sprang directly from leaders of tribal groups, while others, such as the Safavid, came to power through tribal support. Then, as late as 1906 tribal forces influenced the course of political development at Tehran.

Tribal groups and settled communities had their distinctive systems of internal authority. The authority of the tribe, or of a confederacy of related tribes, resided in the *khan*, with below him the *kalantars* who headed each sub-tribe. For generation after generation these leaders came from the same families, and exercised their authority in an acceptable pattern. The khan was at once strong and conciliatory, and blunt and devious, for discipline was maintained by coercion as much as by force. He was hospitable, accessible to everyone, and was the sole source of internal justice which depended more on long established custom than on the shar'ia.

The towns were administered by a number of officials. They were divided into wards or quarters, each headed by a *kadkhuda* who was responsible to the *kalantar* of the town. The civil authority of the kalantar was supplemented by the religious authority of the *mutasib* who was concerned with the enforcement of the shar'ia and of public morality.

Extensive covered bazaars distinguished the towns from the villages, and separate sections of the bazaars were occupied by the combined workshops and retail sales outlets of the *asnaf*, or craft guilds. Apprentices in these crafts were the *shargirds*, while the *ustads*, or master craftsmen, made up the governing power of the guilds. The guilds stressed the hereditary aspects of the crafts, assured high standards of craftsmanship, and provided for mutual cooperation and assistance among its members. Other associations included the *lutis*, the *dashha*, and the *zur khaneh*. The first two of these were guild organizations devoted to the preservation of public morality and security and to the aid of those in distress. Later on, the first two groups tended to degenerate into groups of strong-arm men who preyed on the populace.

The *zur khaneh*, or house of strength, was, and is, a physical cultural society with members from all social levels. It is believed to have originated not long after the Arab invasion of Iran, when young men were trained in secret against the day the invaders could be driven out. A series of exercises are performed to the accompaniment of stirring drum beats and the chanting of verses from the *Shah nama*, with the military origin of the institution attested by the manipulation of heavy wooden shields and bows of iron with strings represented by links of chain.

The farmers lived in some 40,000 villages, most of them so remote from the main roads that they were little affected by the recurrent invasions, and it may well be that the isolation of the villages was a major factor in the preservation and continuity of Iranian civilization. At every period men came from these villages to make their marks in the arts and in positions of power. Bound to the land, living and cultivating the land in the same way for centuries, the situation of the farmers was fairly easy when the monarchy was strong enough to spread the benefits of irrigation and of public safety, but very unfavorable when political weakness led to economic chaos.

Each village was headed by a *kadkhuda*. In early Islamic times, the kadkhuda seems to have been elected, while later he was named by the central authority but was nearly always a local inhabitant. More recently, he was the personally named representative of the landlord who owned the village, including its fields and houses. In more distant times the villages were self-sufficient, self-governing units, paying taxes to the state through the medium of the kadkhuda. Other village functionaries included the *dashtban*, or watchman of the fields, the *mirab*, who regulated the flow of water to the fields, and, less frequently, a bathkeeper and a mulla. In addition, the *rish safid*, the "white beards," or the village elders, contributed their experience to resolve minor intervillage quarrels and rivalries.

The village peasant himself, *ra'iyat*, or *dehgan*, saw little of the country beyond his immediate horizon; he viewed the townspeople with distrust, and had them neatly categorized with uncomplimentary epithets.

Factional strife and community isolation were factors inimical to communal solidarity. In the villages and towns, the Haidari and Ni'mati factions—one taking its name from Sultan Haidar and the other from Shah Ni'matullah, whose shrine is at Mahan, near Kerman—long engaged in bitter rivalry and in conflicts which sometimes resulted in loss of life. In addition, violent outbursts of enmity occurred between neighboring villages, usually over water or grazing rights.

Throughout the centuries, Persian society remained strongly family-oriented. The life of the individual revolved around his family, his relatives, and, frequently, a few close friends. Society was precisely ranked and graded. A manual of administration written in the Safavid period took cognizance of the existence of this social ladder by listing the titles awarded to every occupation or position in the state from that of the shoemaker up through the entourage of the shah. Resounding honorary titles bestowed by the shahs finally evolved into family names.

Related to the use of titles and honorifics was the highly developed practice of *ta'arof*, an Arabic word with the meaning of ceremonial courtesy. It concerned the forms of submissives and honorifics employed among individuals and was an important ingredient in the cement of the social order, since the formulae of address and manner of behavior related to positions on the social ladder. Ta'arof found its full flowering in *tashrifat*, defined in Persian as "honoring" and "ennobling," which covered the areas of ceremonial greetings, ceremonial behavior, and ceremonial courtesy. The degrees of formal courtesy and behavior were in direct relationship to the person's rank on the social ladder. More important, formal courtesy served to protect the individual from the corrosive demands of the world; they served to isolate him from direct involvement in the affairs of others. In being well mannered and in deliberately refraining from causing trouble to others, he played his proper social role.

A number of positive qualities of personality were highly regarded. *Sha'n* denoted the relative rank and dignity of the individual: his "face" within the system of ta'arof. *Gheyrat*, or honor, related to a person's self-respect, his obligation to do his "sacred duty" under appropriate circumstances, while *shaksiyat*, or personality, concerned the personal integrity of the individual. In a proper combination these qualities served to give a man a "good name," implying that he behaved correctly, got on well with others, and was hospitable and generous.

Religion had its impact on social behavior. One example was the practice of *taqiyeh*, mentioned earlier, which may here be redefined as "legitimate dissimulation for religious purposes." This dissimulation for the self-protection of the Shi'a was gradually transferred to many aspects of personal behavior. This self-imposed discretion about personal affairs, became so deep rooted that it gave rise to a skeptical reaction toward things which appeared to be open and straightforward.

Literature and the spoken language both influenced and reflected social behavior. Contrivances of language employed to avoid giving offense to others, to avoid involvement in the affairs of others, and to avoid making firm commitments include ambiguity, vagueness and lack of affirmation. The prevalence of ambiguity in poetry was described earlier, here it may be defined as "the appropriate use of equivocal words or expressions in order to purposely remain vague." Ambiguity features the vocabulary of Persian, with numerous verbs having many meanings some of which are unrelated to the others. Lack of affirmation may be illustrated by the word used for the negative, *na kheyr*: it actually means "not good" or "not beneficial" rather than "no."

On all levels of society above that of the peasants, a contradiction existed between admired virtues and values and the willingness to display these virtues in public life. While greatly respecting in principle such qualities as courage, sincerity, honesty, self-sacrifice, and the ready expression of deeply felt emotions, in practice those individuals who wished to move up the social ladder or to cling to privileged positions behaved with caution, adapted themselves to dishonest conduct, and were careful and reserved in their actions and the expression of their true convictions. Although the display of these theoretical virtues was admired, there was usually a lingering suspicion that apparent virtue must be a mask for more normal behavior. Two factors, already mentioned, lay behind this retreat from conspicuous ideal behavior: the ingrained habit of dissimulation, and the advisability of not attracting the attention of authority.

In summary, Persian society preserved its character through a deliberate emphasis on those patterns of behavior and of social relations which could best withstand upheavals, stresses, and strain, and keep the family aloof from the world around it. How these social patterns have altered in recent years will be examined in the account of modern Iran.

MODERN IRAN

Note: With reference to dates given in this section, a note of explanation is necessary. Dates given in Christian years are approximations of the Iranian year; that is, the Iranian year 1330 is given as 1951, although this year began on March 21, 1951, and ended on March 20, 1952.

IV. THE PAHLAVI PERIOD

O̶N February 21, 1921, just five days before the Irano-Soviet Treaty of Friendship was signed at Moscow, the weak and vacillating government in Tehran was overthrown by a combination of political pressure within the capital and military pressure by the troops who had marched to the city from Qazvin. The leader of these troops was General Reza Khan. Reza Khan was born in 1878 into a local family of landowners and military men at Alasht, a small village high on the northern slopes of the Alborz range. His father died before his son was a year old and the young mother returned to Tehran, her former home, with the baby. Friends and relatives helped the growing boy, saw that he learned to read, and enrolled him in the Cossack Brigade when he was sixteen. This brigade, later a division, had been established earlier following an Iranian request to the Russian Czar to help form a force modeled upon the Russian Cossacks. Staffed in part by Russian officers, it was the best trained force in the country.

Over a period of many years, Reza Khan advanced from rank to rank. A man of impressive stature and great reserve, he was renowned for bravery in numerous conflicts. A man with few close friends and no confidants, he burned with a consuming passion to rescue Iran from "the villainy of foreigners and the treachery of mean Iranians." It was his belief that the armed forces had been disgraced by the governments of Iran, and that it was up to the army to rejuvenate the country.

By 1921, he had risen to the rank of general in the Cossack Division. After having tasted defeat at the hands of the Soviet-organized forces in Gilan, he was retraining his ill-equipped men at Qazvin, some 60 miles to the west of Tehran. Suddenly, chance plunged him into the political arena, but it seems probable that in time he would have made his own move to rise to power.

At Tehran, Sayyid Zia al-Din Tabatabai, a crusading journalist, was preparing for a *coup d'état* which was to overthrow an ineffectual cabinet and install a government which would combat foreign influences and institute reforms based on Western models. Needing a military force to occupy Tehran, he sent emissaries to Reza Khan who responded with enthusiasm.

Ahmad Shah named Tabatabai as prime minister, and Reza Khan became commander-in-chief of the armed forces. Confronted by the opposition of entrenched social forces, Tabatabai was unable to carry out his program of reforms, and after 100 days resigned and left Iran. Reza Khan was Minister of War in several successive cabinets, until in October 1923 Ahmad Shah appointed him prime minister and left for Europe for "reasons of health"—never to return. Within a year after he moved into this post, some of Reza Khan's advisors were advocating the establishment of a republic, pointing to the success of neighboring Turkey. He approved of a campaign which was undertaken throughout the country in favor of a republic, but this effort ran into opposition of the Majlis and the Muslim religious leaders. Reza Khan met with these leaders and called a halt to the campaign.

In 1925, Reza Khan initiated the first of a long series of actions designed to promote national unity. Actually, they established conformity rather than unity—which is not responsive to decrees. In May he eliminated the use of honorary military titles and led the way in a move to adopt family names by taking that of Pahlavi. In that same year a special Constituent Assembly chose Reza Shah Pahlavi as the first ruler of the new dynasty. He was crowned in the spring of 1926, with his first son, Shapur Muhammad, then six years old, standing beside the throne.

Reza Shah felt more keenly than any of his compatriots the tragic contrast between Iran's glorious past and her present impotent state, and was resolved to rouse the country from her lethargy and to foster national unity and pride. Iran was to throw off all foreign intervention and influence and to win full independence and the respect of other nations. She was to be industrialized, and her social and economic institutions reformed, along Western lines, a program similar in working details to that of neighboring Turkey. Under it remarkable progress was made during the next few years.

In the field of foreign affairs Reza Shah terminated the system of capitulations, negotiated customs autonomy, and put an end to the practice of seeking loans abroad. British interference in Iranian affairs was no longer apparent; the right of currency issue was taken away from the British-owned Imperial Bank of Iran, and the contract with the Anglo-Iranian Oil Company was canceled and replaced with one more advantageous to Iran. Other steps were directed at foreign interests and influence. Foreigners were prohibited from owning land, and restrictions were placed upon foreign-owned shops and businesses. Members of the foreign service were forbidden to marry foreigners without express permission, and officials of the government were directed not to attend functions at the legations, and to avoid social and other contacts with foreigners. It was decreed that all other countries must call the nation "Iran" rather than "Persia," and mail addressed to Persia was returned to the sender. Reza Shah said that Iran must learn to do without foreigners, and that after a very few years foreign specialists and advisors would no longer be required. He added that for too long his people had relied upon others, and their characters must be hardened, so that they could become independent in thought and action.

Reza Shah displayed implacable hostility to local manifestations of communism and jailed both socialistically inclined and Soviet-inspired labor leaders, writers, and teachers. In 1931 the Majlis enacted a bill which was designed to deal with individuals spreading foreign ideology. It provided prison terms for those advocating forcible overthrow of the political, economic and social order, for those endeavoring to separate territory from Iran, and for those trying to weaken those patriotic feelings which related to the independence and unity of the country. At the same time the ruler realized the value of maintaining close commercial and economic ties with Soviet Russia. In 1927 five pacts, and in 1935 several more, were concluded between Russia and Iran, covering fishing rights in the Caspian, the return to Iran of installations at the Caspian port of Pahlavi, commercial relations, customs tariffs, and guarantees of mutual neutrality and security.

Reza Shah created a large army, provided with modern

weapons, which soon brought every corner of the land under the control of the highly centralized government. An age-old social order was sacrificed to the new progressive regime. Nobles lost power and prestige. The merchants lost freedom of enterprise as they were drawn into the governmental system of monopolies and of controls over industry, commerce, and trade. The tight grip of the Muslim clergy over many phases of public life was a challenge to the position of the new ruler, who took steps to break down their power and prestige. The clergy lost direct control of much of its vast trust funds; religious law gave way to civil and criminal codes; licenses were required for the wearing of clerical garb; civil marriage and divorce registers were established; non-Muslim foreigners were permitted to visit the splendid mosques of the country; religious passion plays were suppressed; dervishes were forbidden to appear in towns; and religious teaching gave way to state schools. The object of direct attack was not religion itself, as in certain other countries engaged in a reworking of their social structure, but rather those forms and expressions which were clearly outmoded in a revitalized Iran.

Reza Shah brought women into society. Not only did he remove their veils and introduce them to Western clothing, but he insisted that they play an active role. He said, "Up to this time one half the population was not taken into account, and there were no statistics of the female population; it seemed as if women were some other type of individuals who did not form a part of the population of Iran. Now that you have entered society, it is your duty to work: you are to be the educators of the next generation."

Measures aimed at building national unity by imposing conformity continued. The men of the country were told to wear the Pahlavi hat, a visored headgear similar to that of a French army officer. Later this was replaced by the European felt hat. The Majlis passed a "uniform dress law," which made the wearing of Western clothes compulsory, thus replacing the many different tribal and regional costumes. Another step towards uniformity, as regarded social status, was the abolition of all honorific titles, as well as the special titles employed by tribal chiefs, large landowners, and relatives of the earlier ruling dynasty. Only terms corresponding to *mister*, *madame*, and *miss* were permitted.

Other specific actions taken to promote uniformity, such as the establishment of a single system of civil courts, the opening of notary offices for the recording of vital statistics, contracts, and land ownership, and the introduction of the metric system in place of widely differing regional weights and measures, were not specially related to the larger goal of nationalism.

Efforts to modernize and industrialize the country were quite stupendous in light of its limited financial resources, and the fact that Reza Shah refused to consider foreign loans. The Trans-Iranian railway was financed entirely by special taxes levied upon tea and sugar. An extensive network of graveled roads was built; trucks and passenger cars imported; sugar refineries, cement plants, cotton, silk, and woolen textile mills erected; the growing of tea was promoted; and a number of state monopolies concerned with major items of export and import managed to maintain a generally favorable balance of trade. The ruler insisted that Iran must increase its exports of raw materials in order to obtain the foreign exchange required for industrialization. Unfortunately, in spite of numerous warnings, the landowners failed to take the steps necessary to improve irrigation facilities and to expand the areas devoted to the growth of staple foods: there was actually a decline in agricultural productivity.

Throughout his reign, Reza Shah was concerned with education, believing that in universal education lay the key to national strength and unity. Schools all over the land taught Persian rather than the local tongues of some of the regions, and new textbooks and readers were designed to instill the same patterns of social behavior in all areas. In addition, Reza Shah was the initiator of the reform of the Persian language by which it was purged of words taken from other languages.

Reza Shah was twice outside of Iran, once when he went to the Shi'a shrines in Iraq, and in 1934 when he visited Ataturk. Touring parts of Turkey together, the two got along famously, in part because Reza Shah was fluent in Turkish.

He had thought of going on to Europe, but was depressed by the fact that Turkey's progress seemed greater than that of Iran, and he returned directly home, determined to make his people work harder.

He did not fit the traditional stereotype of a head of state who plays off conflicting interests against each other, shows favoritism to powerful elements, and seeks security through inaction. In spite of his conspicuous faults—his acquisition of property and wealth, his abuse of power, and his persecution and destruction of individuals whom he believed were plotting against him—he had clear insight into the nature of the Persians and knew the ingredient most essential for economic progress. This ingredient was work: he was consumed by a determination to put an end to the climate of lethargy, and to make the people work hard. He himself had no hobbies and took no vacations and was happiest when he opened new schools, hospitals, and factories. One day he asked the cabinet ministers to say what they regarded as his most important contribution to the nation's progress. One said the railroad, another compulsory education, and still another, the industrial development of the country. He replied that they were all mistaken: his greatest contribution had been to make the people aware that in work lay the bright future of Iran. He said: "I have made the Iranians realize that when they get up in the morning they must go to work and work hard all day long."

Reza Shah failed to establish effective channels through which his national goals could be made clear to the people, and which would serve to draw them into active participation to achieve these objectives. Four political parties were briefly active in 1927, and word got around that the ruler planned to sponsor a majority political party to support the programs of the government within and outside of the Majlis. In 1932 he raised the subject with a group of deputies, informing them that a Majlis without political parties was defective, and that a patriotic political party should be formed which would press for further reforms. Nothing came of this proposal: his authoritarian attitude would have not welcomed political challenge, while in 1909 he had witnessed the difficulties caused to the parliamentary system by a multiplicity of parties.

As a possible channel of communication, in January 1939 the Society to Guide Public Opinion was established. Influenced by the Nazi and Fascist propaganda machines, the society was to direct public opinion along specific lines, to inform the public about the plans of the government, and to promote national pride. Lecturers toured the cities, and a flood of publications appeared. Some of the material stressed that individuals should subordinate their personal interest to the higher ones of the state, and that the intellectual elite must give its fullest service to the nation.

On September 4, 1939, Iran formally declared its neutrality in the European conflict and hoped to carry on normal relations with all the powers. A few weeks later it protested a decision by the British government to prevent shipments from Germany to neutral countries. Germany had become Iran's largest trading partner, and Iran had orders in for vitally needed machinery, construction materials, and manufactured items. In the following year the situation appeared to have been eased when Iran and the USSR reached an agreement providing for the free transit of goods shipped through the Soviet Union. Even earlier Reza Shah had broken a cardinal principle when he instructed his officials to seek loans from the United States for the purchase of railway equipment and military supplies. Then Great Britain, following the collapse of France, cancelled all previous orders from Iran for similar materials. In 1940 Iran successfully pressed the Anglo-Iranian Oil Company for higher royalty and other payments: this action aroused the personal enmity of members of the British government towards Reza Shah.

On June 22, 1941, Germany invaded the Soviet Union, and four days later Iran declared its neutrality in this new conflict. In Iran the feeling prevailed that Germany would win the war. Wishful thinking or not, it was thought that Germany would mount a successful winter drive through the Caucasus region and that the final German victory would result in the return, to a friendly Iran, of those areas of the Caucasus and Turkmanistan taken from Iran by the Russian empire. Russia and Great Britain, newly allied, were also aware that a German drive to Iran through the Caucasus would menace the Soviet rear. Iran was also important as a protected supply route to Russia, and the British navy was dependent upon the output of the oil fields of the country. Germany, equally aware of Iran's importance, had a built-in channel of communications with the local elite and officials through its nearly 700 citizens who were employed by the Iranian government or by German firms in the country, while its secret agents were active in organizing centers of support among the tribes of the southwest.

The Allies made oral and written demands that most of the Germans be expelled from Iran. That government replied that such an action would violate its neutral posture, but later stated that its policy was to send away all foreigners whose jobs could be filled by Iranians, and that this process was proceeding rapidly. Reza Shah felt that the Allied notes must conceal some more serious demand, but he could not find out what it was. Actually, the increasingly forceful tone of the notes was a tactic to provide a rational excuse for a pre-determined invasion of Iran. The writings of Winston Churchill and others make it clear that the invasion was de-termined on well before the event.

On August 26, 1941, Russian forces entered Iran from the northwest and British troops marched across the Iraq frontier and also landed at the head of the Persian Gulf. British ships staged a surprise attack upon the Iranian naval forces at Khor-ramshahr, sinking every vessel with a considerable loss of life. The Iranian army put up a token resistance which was called off in three days.

On August 30 the Allies addressed a memorandum to the Iranian government which contained a number of demands: all German and Italian citizens to be expelled within one week, diplomatic relations with Germany and Italy to be bro-ken off, Iran must refrain from any hostile acts against the Allies, Iran must facilitate the transport of Allied war mate-rials by road, rail, and air, and Iranian forces must retreat from the areas occupied by the Allied troops. Early in September Iran accepted these demands, and signed an agreement which placed the greater part of the country under the control of the British and Soviet forces. The government was informed that a joint Anglo-Soviet occupation of Tehran would take place in the near future. Throughout these days Reza Shah had remained silent; now he was attacked in bitter terms by British radio stations in London and Delhi. Although he would have suffered any indignity to remain in Iran, the ruler realized that the Allies would be unwilling to make any accommodation with him, and that he must act promptly in order to secure the succession for his son.

On September 16 he left Tehran for Isfahan, and from that city sent back a letter of abdication stating that the entire nation should recognize his heir and legal successor as its sovereign. The next day Shapur Muhammad, now Muhammad Reza Shah, took the oath required of the sovereign. A few days later the former ruler transferred all his property to his successor in return for ten grams of lump sugar, requesting that its value and revenues be spent for works of charity, education, and the like. Taken in charge, along with several members of his family, by the British, Reza Shah was shipped off to the island of Mauritius and then to Johannesburg in South Africa. From the moment he left Iran he lost the will to live, for he loved Iran more than anything else, more than his family and more than life. In July 1944 he died of a broken heart.

Born on October 26, 1919, Muhammad Reza Pahlavi was the first son of Reza Shah, and the twin of a daughter, Ashraf. From early childhood his father trained him, with constant affection. He first attended a special school in company with some of his brothers and a few other boys. In 1931 he was sent to Switzerland, spending three years at Le Rosey boarding school where he did well in his studies and was active in sports. In 1936 he entered the Military College at Tehran, and two years later was graduated first in his class. From this time on he accompanied his father on many tours of inspection throughout the country. Reza Shah would explain his plans and ask for his son's opinions, and often there were discussions of patterns of Iranian behavior. In 1939 he married Fawzia, a daughter of King Fuad of Egypt. A daughter was born of this union which later ended in divorce. The years of Muhammad Reza Pahlavi's sovereignty fall into three periods. First, the years of World War II when he was still a very young man and Iran was dominated by the Allied forces. Second, that from the end of the war until August 1953 when he reigned; and, third, from that latter date when he ruled. It may be thought that ruling did not come easily to him: his democratic schooling in Europe led him to question the dictatorial methods of his father, and, in addition, he was not aggressive by nature.

Once on the throne, his first Prime Minister was Foroughi, a philosopher, author, and respected elder statesman, who devoted his entire attention to the conclusion of a friendly working agreement with the Allies which would favor the prosecution of the war against Germany. On January 29, 1942, the Tripartite Treaty of Alliance was signed between Iran, Great Britain, and Soviet Russia. In the first article of the treaty the two Allied powers undertook to respect the territorial integrity, sovereignty, and political independence of Iran. The Allies agreed to use their best endeavors to safeguard the economic existence of the Iranian people against privations and difficulties arising as a result of the war. Other articles specified the facilities to be granted the Allies for the struggle against Germany, but stated that the presence of Allied forces on Iranian soil did not constitute a military occupation, and that all such forces would be withdrawn from Iranian territory not later than six months after all hostilities between the Allied Powers and Germany and her associates had been terminated.

Military supplies destined for Russia began to arrive at the head of the Persian Gulf early in 1942, to move up over the Trans-Iranian railway and the motor roads leading north. In December 1942 the first American troops arrived in Iran; a force, known as the Persian Gulf Command, which eventually numbered 30,000 men. The Americans operated the rail line from the Gulf to Tehran and maintained a truck convoy system from the Gulf to the north of Iran where the supplies were turned over to the Russians. The truck line transported 400,000 tons of war materials, while the railroad carried over four million tons of supplies, for Soviet Russia and Iran. Working together as an efficient and cooperative team, the Allies moved over five million tons of war materials across Iran and into the hands of the Russian army.

By the spring of 1942 diplomatic relations with Germany, Italy, and Japan had been severed and their nationals expelled from Iran. On September 9, 1943, Iran declared war on Germany, and announced her adherence to the Declaration of the United Nations.

At the end of November 1943 Roosevelt, Churchill, and Stalin arrived in Iran for the Tehran Conference. Of vital moment for the Iranians was the Tehran Declaration, in which the Allied leaders recognized the aid given by Iran in the war, agreed to continue economic assistance to Iran, stated that economic problems confronting Iran at the end of the war would be given full consideration by the proper international conferences or agencies, and expressed their desire for the maintenance of the independence, sovereignty, and territorial integrity of Iran.

The impact of the war upon Iran was severe. Food was short and goods of every type very scarce, inflation sent prices soaring, and large amounts of currency issued to meet Allied expenditures within the country added to the inflationary trend. Successive short-lived cabinets were too concerned with daily emergencies to be able to consider the fundamental problems resulting from the swift changeover from the autocratic reign of Reza Shah to the more democratic form of government. The members of the Majlis elected in 1943 failed, due to lack of unity and purpose, to give sustained support to successive cabinets. Prime Ministers followed each other at frequent intervals. Successive cabinets were charged with favoring the interests of foreign nations, with being too conservative, or with being too radical.

In 1944 American and British oil companies made overtures for concessions in southeastern Iran, and in October of the same year Soviet Russia requested an oil concession in the north of the country. The answer of the Majlis was to pass a bill prohibiting any Iranian official from concluding an agreement with any foreign power or interest regarding any concession. Iranian leaders decided that decisions vital to the interest of the nation should be made only after the end of the war, and in October 1945 the Majlis passed a bill which postponed elections for the next session of that body until six months after all foreign troops had left the country.

The new political parties were extremely vocal. In the province of Azerbaijan the radical *Tudeh*, or "Masses," party was nominally replaced by the so-called Democratic Party of Azerbaijan. This group, whose field of activity was occupied by Soviet troops, vigorously denounced the policies of the central government and attacked Iranian army garrisons and police posts throughout the province. The area was in open disorder, and in November 1945 government reinforcements were sent from Tehran, only to be halted by Soviet forces stationed at Sharifabad, to the west of Tehran. In December the Democratic Party of Azerbaijan, announced the establishment of an autonomous state of Azerbaijan, and at the same time the Russians set up another puppet state, the Kurdish Republic of Mahabad, also in Azerbaijan. In January 1946 the Iranian government notified the Security Council of the United Nations of alleged interference by Soviet Russia, through the medium of its officials and armed forces, in the internal affairs of Iran. The Council unanimously adopted a resolution stating that Iran and Russia should inform the Security Council of the results of their negotiations on this matter, and retained the right to request information on the progress of the negotiations.

In February 1946 Ahmad Qavam became prime minister of Iran and went at once by plane to Moscow for negotiations. He was in that city on March 2, the date previously agreed upon by the Allied powers for the complete evacuation of their troops from Iran. The Americans had left earlier, and the British forces moved out by the final date, but Soviet Russia announced that the evacuation of its forces from some areas of northeastern Iran would begin on March 2, while troops in other areas would remain pending clarification of the situation. On March 18 Iran protested to the Security

Council that the USSR was maintaining troops in Iran territory contrary to the Tripartite Treaty, and that the agents, officials, and armed forces of the USSR continued to interfere in the internal affairs of Iran. The question was discussed by the Council. Near the end of the month talks between Ahmad Qavam and members of the Soviet Embassy at Tehran resulted in an exchange of notes, made public on April 4, which provided for the withdrawal of all Soviet forces from Iran within a period of six weeks; for the presentation to the next session of the Majlis of a proposal for the formation of a joint Irano-Soviet oil company which would exploit the oil resources of the northern provinces of Iran; and for the peaceful internal settlement of differences between the central government and the revolutionary movement in Azerbaijan. The agreement regarding the withdrawal of Soviet troops was reported to the Security Council, which adopted a resolution to defer further proceedings in the Iranian question until May 6, at which time both the USSR and Iran were to report to the Council whether the withdrawal of all Soviet troops had been completed. On May 22 the Council suspended discussion of the Iranian question. The Soviet troops finally evacuated Iran.

Prime Minister Qavam then turned his attention to internal affairs and prepared to deal with the open conflict between the right and left wing political parties, with tribal unrest, with the problem of Azerbaijan, and with basic reforms.

Sayyid Zia al-Din Tabatabai had returned earlier to Tehran, having lived outside of Iran since 1921. He headed the National Will party which flourished until his arrest, and then vanished from the scene. The prime minister announced the formation of his personally sponsored party, the Democrats of Iran. In the summer of 1946 he accepted the collaboration of the Tudeh party, and three of its founders, who were generally regarded as Communists, were taken into his cabinet. In September the penetration of Tudeh elements into southern Iran and dissatisfaction with the too conciliatory attitude of the government with regard to the Tudeh leaders and the Azerbaijan regime, culminated in open revolt by southern tribes. Close cooperation between the Qashqa'i,

Mamasani, Hayat Daudi, and Tangestani tribes resulted in the seizure of the towns of Kazerun, Bushire, and Bandar Dailam. The government realized that it would lose support in other parts of the country as well unless it displayed a harder, nationalistic attitude. The cabinet was re-formed without the members of the Tudeh party, and the Democrats of Iran seemed to dominate the political picture.

Long-drawn-out negotiations were held with the leaders of the so-called government of Azerbaijan and general points of agreement reached. In October the Shah issued a royal decree providing for elections for the new Majlis, and the prime minister announced that the elections would be held throughout the country as soon as the security forces of the nation were everywhere in position to supervise the voting. The insurgent government of Azerbaijan now went back on its agreement to allow such supervision. Therefore, early in December 1946 the Shah himself surveyed the insurgent defense positions from the air, and then ordered the regular forces to move on Tabriz. The city was quickly taken, both puppet states disappeared, and scores of Soviet-trained agents and local communists fled across the Russian frontier.

The recovery of Azerbaijan was celebrated throughout the country, and public recognition of the ruler's bold decision appeared to have encouraged him to undertake a broader interpretation of his constitutional powers. Over a period of time he sought to play a more active role in relation to the Majlis. The Shah was well aware that his father had selected the deputies to be elected and that the government still disposed of the means, through the Ministry of the Interior, to influence the results of elections: officials began to pass on word of his choices. Reza Shah had chosen his prime ministers who, along with their cabinets, were then confirmed by the Majlis. After his abdication the Majlis reverted to the method in effect prior to 1926, a system in which the deputies discussed the merits of various candidates, voted on them, and sent the name of the man receiving most votes to the Shah for his royal appointment. Now Muhammad Reza began to indicate his choice for the next prime minister, and, in 1950, went a step further by naming a prime minister without consulting the Majlis: this procedure remains in effect.

He pressed for the convening of the Senate, a chamber which had been provided for in the constitution, with legislative powers similar to those of the Majlis, but which had never been formed. In 1949 the Senate was established: half of the sixty-man body is appointed by the Shah, and the rest by elections in two stages. Allegedly more conservative than the Majlis, its right of veto over bills enacted by the Majlis serves to protect the nation from legislation "inappropriate" for Iran. In 1957 an amendment to the constitution provided for an increase in the number of deputies in the Majlis and extended its terms from two to four years. It also gave the ruler power to withhold his assent to bills passed by both houses, although they would become law if again approved by three-quarters of the membership of both houses. In addition, the amendment gave the Shah the power to dissolve both the Majlis and the Senate, after giving his reasons and then issuing a call for new elections.

In presenting this outline of the Shah's actions in relation to the Parliament, the chronology of events has been neglected. The details of this chronology may be found in earlier editions of this work: the present approach is to stress the most significant events.

The XVth Majlis was convened in July 1947, and with most of the deputies members of the Democrats of Iran party, Qavam was again chosen prime minister. On October 22 Qavam presented that body with a report on the negotiations leading to the April 1946 agreement for the establishment of a joint Irano-Soviet oil company. The Majlis at once passed, by a vote of 100 to 2, a bill containing several clauses. These voided the 1946 agreement as contrary to a law of 1944 forbidding such negotiations by government officials; provided that experts were to survey the oil resources of Iran during the next five years and that if oil were located in the north in commercial quantities the government could enter into negotiations for the sale of such oil to the USSR; stated that in the future no concessions would be granted to foreigners or to companies in which foreigners had any interests; and, directed the government to study concessions granted to foreign companies, especially the southern oil concession, with a view to increasing the benefits derived by Iran from such concessions.

After this date opposition to Qavam grew among the Democrats of Iran, in the Majlis, and within his own cabinet, and on December 10 Qavam resigned after failing to win a vote of confidence from the Majlis. From this date until mid-1950 there were five different cabinets. Neither the prime ministers nor the Majlis displayed significant activity, and debate and procrastination were substitutes for decisions. In February 1949, a would-be assassin fired five revolver shots at the Shah at point-blank range. One bullet inflicted a very slight wound to his face, the others tore through his uniform. Escaping almost certain death, Muhammad Reza felt that his preservation represented the second in a series of miracles, the first of which had been the recovery of Azerbaijan, and he became convinced that he was under divine protection to accomplish a mission for his country. The Tudeh Party of Iran, held responsible for the outrage, was banned and those of its leaders who did not flee the country were tried and sent to prison: this action did much to enhance internal tranquility and security.

During 1949 the Majlis approved the program of the Seven Year Development Plan and authorized the establishment of a Plan Organization to execute its projects. The record of economic development under this and later plans is considered in another section. Near the end of the year the Shah made his first state visit to the United States. Now speaking fluent English, Muhammad Reza pressed for economic and military aid. In this first effort he was unsuccessful. However, less than a year later the United States announced that the first Point IV program to be carried out in any country would be initiated in Iran, and that arms would be sent to Iran in recognition of that country's adherence to the principles of the Mutual Defense Aid Program.

In June 1950 the Shah named General 'Ali Razmara as prime minister. An able military administrator who had been chief of staff of the armed forces, this "tactician"—the meaning of his name—was quickly involved in the issue of the future of Iran's oil, and in the opening phase of a bitter struggle between the Shah and his supporters and Dr. Muhammad Mossadeq and his National Front.

Razmara approached the oil issue by requesting the Majlis to take definite action of some kind on the still unratified Gass-Golshayan agreement of July 1949, an agreement which would have about doubled the royalty payments made by the Anglo-Iranian Oil Company. At first his position on this agreement was a neutral one, then he asked for favorable action, and finally he withdrew the proposed agreement from the Majlis after its Oil Committee had voted unanimously against acceptance.

In March 1951 events took a violent turn. On the third of that month Razmara appeared before the Majlis to present a report prepared by his government which stressed the fact that Iran still lacked the number of skilled technicians required to operate the oil industry and he went on to indicate that the time was not ripe for the nationalization of the industry. On the seventh he was shot and killed while attending a ceremony in a mosque. His assassin, Khalil Tahmesbi, a member of the militant religious group called the *Fedayan-i-Islam*, or Devotees of Islam, was arrested, and two days later his organization threatened to kill the Shah and other leading figures if he were not released.

Husayn 'Ala, long a respected and experienced diplomat and public official, took over as prime minister to head a government whose task was soon complicated by the fact that the Majlis seemed resolved to act independently of the government. On March 15 the Majlis passed the bill providing for the nationalization of the Anglo-Iranian Oil Company, and on March 20 the Senate approved the same bill. Demonstrations and rioting broke out, with the Tudeh party demanding the immediate seizure of the oil installations. Three British nationals were killed in rioting at Abadan and the United Kingdom reacted to this disturbing situation by announcing that it was "ready to act as we see fit to protect British lives and property." Deputies reproached 'Ala for his alleged unwillingness to implement nationalization and, as the public followed their lead, his position became untenable.

Following a unanimous vote by the Majlis for the immediate seizure of the oil industry, on April 29 Dr. Muhammad Mossadeq was named as prime minister. He had been active as government official, deputy, and cabinet minister in the years immediately following World War I. In 1925, as a member of the Majlis, he opposed the elevation of Reza Khan to the throne, and paid for this rashness by a period of detention and then years of enforced residence in a quiet village. After the Allied occupation of Iran in 1941 he reappeared at Tehran and was chosen from the capital as a deputy to the XIVth Majlis, owing his election in part to his reputation for sincerity, deeply felt convictions, and personal honesty. In this Majlis he made a number of long speeches which hammered away at a single theme: that Iran must free itself from foreign influence and foreign domination. In a notable speech begun on October 29, 1944, and carried over for two days, he dealt at length with the history of oil concessions in Iran, and made a classic attack upon the position of the Anglo-Iranian Oil Company. Near the end of this speech he stated: "The Iranian nation wants a political equilibrium which will be in the interests of this country and that will be a 'negative equilibrium.'" While not precisely defining "negative equilibrium," he implied that in the past Iran had survived the pressure of competing great powers by first giving something to one and then something else to another: this was positive equilibrium. He advocated the opposite course, the course of granting no favors or concessions to any powers and hence giving none of them grounds for complaining that his interests had been ignored at the expense of those of a rival. In December 1944 Mossadeq was instrumental in drafting a bill passed by the Majlis which prohibited any official from concluding an agreement with any foreign interest or power regarding a concession in Iran. Again, in October 1947 Mossadeq led the Majlis in the passage of a bill which rejected the proposal of the government that a joint Irano-Soviet oil company be set up to exploit the oil of northern Iran. The same bill contained an article directing the government to study the southern oil concession (of the Anglo-Iranian Oil Company) with a view to increasing the benefits derived by Iran. In this article was the germ of the nationalization law, and in steadily pushing for the complete elimination of foreign participation in Iranian enterprise Mossadeq was transformed from a rather uninspiring reciter of facts and figures into a leader of public opinion. In the Majlis and other areas of political activity Mossadeq was variously supported by the National Front, a loose grouping of diverse elements—the Iran Party, the Toilers' Party, the neo-Nazi Sumka Party, the ultra-nationalist Pan-Iran Party, the religious fanatics of the Devotees of Islam, and the rabble-rousing religious leader, Sayyid Abol Qasem Kashani. In time the National Front was to fall apart, as most of its elements turned against Dr. Mossadeq.

As prime minister Mossadeq had two goals, each with two parts. First, to win international approval for the nationalization of oil, and to obtain impressive financial returns from Iran's exploitation of this resource. Second—the result of a nurtured hatred of Reza Shah—to undermine the royal powers and to encourage public antipathy towards the Shah, and then to establish a republic. As he moved towards these goals, his own methods became increasingly authoritarian.

The long drawn-out negotiations relating to the nationalization of oil may be summarized briefly. The Anglo-Iranian Oil Company asked the International Court of Justice to arbitrate the issue, and Great Britain brought the matter before the Security Council. The International Bank came forward to offer its services as a friendly intermediary. The United States government took part in attempts to reach a solution and on August 30, 1952, Truman and Churchill sent a joint note to Iran which outlined a broad formula for the settlement of the dispute. This proposal was rejected by Mossadeq, who asked the Anglo-Iranian Oil Company to pay 49 million pounds sterling and then send a mission to Iran for negotiations which could be quickly concluded. When this suggestion was declined, Iran severed diplomatic relations with Great Britain. Actually, the several proposals, made and modified, by the International Bank were favorable to the interests of Iran but did not give Iran a free hand in its nationalization of the oil industry. In May 1953, with the export of oil limited to the occasional loading of a foreign tanker, Mossadeq addressed a letter to President Eisenhower in which he stated that a dangerous situation existed in Iran, and renewed his earlier request for a large loan from the United States. The "dangerous situation" was the implication that unless Mossadeq had funds to prevent financial collapse, communism might take over Iran. On June 29 Eisenhower replied, stating that failing a settlement of the oil issue, the "United States is not presently in a position to extend more aid to Iran or to purchase Iranian oil." However, the letter added that the current technical assistance and military aid would be continued.

Prime Minister Mossadeq began to attack the royal prerogatives in July 1952 when he assumed the post of Minister of War, later Minister of National Defense, as previously this minister had always been selected by the Shah. To weaken the potential of the armed forces for possible action against him, he made changes in the high military command, retired officers, moved army units about, had the Guards Division at Tehran split into independent units, and abolished the traditional ceremony of swearing allegiance to the Shah. These same measures were intended to insure that the armed forces would be directly responsible to him rather than to their commander-in-chief, the Shah. On February 28 word spread throughout Tehran that the prime minister had asked the Shah to leave the country for several weeks. Supporters of the ruler stormed Mossadeq's residence; clad in pajamas Mossadeq fled over the garden wall and went on, still in pajamas, to recount his experience to the Majlis.

On March 12 a special committee of Majlis deputies, chosen to report on rumors of dissension between the Imperial Court and the head of the government, stated that such dissension did not exist and added that according to the constitution the ministers of the government, rather than the Shah, were responsible for the conduct of affairs. Supporters of the Shah felt that the report might be used to remove the army entirely from the ruler's influence, but in spite of Mossadeq's insistence the Majlis failed to act upon the report. At the same time the Tudeh party benefited by the lack of interference on the part of the government—as exemplified by the March ruling of a Tehran court that its members could not be prosecuted for activity against the constitutional monarchy and in favor of communist doctrines—to muster its strength. In April Mossadeq forced the resignation of the Minister of the Imperial Court, and named a successor who was loyal to him and personally distasteful to the ruler.

Although Dr. Mossadeq had announced that his government would bring in public, democratic freedoms, the passing months found him resorting to authoritarian measures. Exercising plenary powers awarded him by the Majlis in 1952—powers which enabled his government to enact laws prior to their approval by the Majlis—he instituted a number of types of control: extension of martial law; imposition of a stringent public security law; prohibition of strikes by government employees and public service workers; suspension of the Senate; suspension of elections for the Majlis; the decree of a very strict press law and the arrest of newspaper editors; dissolution of the Supreme Court; arbitrary retirement of government officials; and the restrictions on movements of citizens and foreigners within the oil-producing area. While some such actions had been taken under previous prime ministers, this activity seemed contrary to the nature of the man who long insisted that elections must be completely free and who had sternly opposed the imposition of martial law and of restrictive press laws.

By the end of June 1953 many groups and individuals had withdrawn from the National Front. Makki, the hero of nationalization who had turned the valve shutting off the flow of oil at Abadan, had deserted Mossadeq as had Dr. Baghai, leader of the Toilers' party. The neo-Nazi Sumka party was in opposition as were elements of the ultra-national Pan-Iranist party. Kashani and a number of other religious leaders of varying degrees of sanctity now condemned Mossadeq at every turn. In spite of the press restrictions a majority of the Tehran papers attacked Dr. Mossadeq, while within the Majlis more and more deputies spoke against his government. Lack of confidence among the merchants was reflected in the fact that the free rate of the rial rose from 75 to the dollar in 1951 to 130 rials in July 1953.

What proved to be the final stage of the regime began with the decision of Dr. Mossadeq to do away with the Majlis. For some months government had been by plenary decree with the Majlis inactive, either because supporters of Mossadeq stayed away to prevent a quorum or because his opponents resorted to the same tactics. Mossadeq resolved to get rid of the Majlis and as an initial step arranged for more than 35 National Front deputies to resign—thus removing the possibility of any quorum. Then on July 25 a decree was issued providing for a referendum on the question of whether the XVIIth Majlis should be dissolved or retained. Voting in conspicuously separate booths, some 2,043,389 were said to have voted for dissolution and only 1,207 for retention. Now Dr. Mossadeq was faced with the difficulties relative to a new election. The previous elections had brought in a Majlis which he had welcomed as being 80 percent representative of the will of the people and then later had denounced. Should he now permit the populace to vote freely, the Tudeh party might elect deputies who would oppose him at every turn. The huge turnout for a Tudeh demonstration on July 20 was a sign of the times. Should he ask the Shah, who alone had such authority, to call for new elections, and, if so, would the ruler who alone had the constitutional right to dissolve the Majlis and the Senate, respond? Some of his associates, and notably Husayn Fatemi, his Minister of Foreign Affairs, spread the word that a second referendum would be held on the question of whether or not the monarchy should be replaced by a republic—a republic with Dr. Mossadeq as its president.

On August 13, the Shah, then in residence at Ramsar on the Caspian coast, issued two imperial decrees: one dismissed Dr. Mossadeq and the other named General Fazlullah Zahedi, then in hiding after having been earlier arrested and released by Mossadeq, as prime minister. In the early hours of August 16 an officer of the Imperial Guards presented the decree of dismissal at the house of Dr. Mossadeq and obtained his receipt of its delivery, while several members of his government

16. Village life: the threshing floor with wheat from sheaves to grain

17. Urban life: active and passive bureaucrats. The *zur khaneh* in the National Bank of Iran

18. Village life: children are seen but not hea

19. Village life: weaving a Persian rug from a full scale pattern

and of the National Front were briefly detained by detachments of the Imperial Guard. On the morning of the 16th, Radio Tehran broadcast a brief item on the collapse of an attempted military *coup d'état*, while Mossadaq, meeting with his cabinet and issuing a decree dissolving the Majlis, said nothing to them about his dismissal. That afternoon a mass meeting arranged by the National Front received word that the Shah and his wife had gone to Baghdad by air and the fervent orators, including Fatemi, were hailed with cries of "Down with the dynasty," and "Death to the Shah" from elements of the Iran party, the Third Force, and the Tudeh party. Participants left the meeting to stream along the streets, smashing showcases and windows housing pictures of the Shah.

On August 17 Fatemi released the text of his instructions to the Iranian embassy in Iraq to avoid contact with the person who had fled, but by that time the Shah had been welcomed by the government of Iraq. In his own paper, *Bakhtar Emruz*, Fatemi poured out abuse against the ruler, comparing him with a snake and stating that "the people of this country want to see this traitor hanged." At Tehran crowds shouting "Mossadeq is victorious" pulled down statues of the ruler and his father, Reza Shah, while the underground central committee of the Tudeh party called for the uprooting of the monarchy and the establishment of a democratic republic.

On the 18th the atmosphere altered slightly as another side of the story began to unfold. Opposition papers managed to appear: they printed reproductions of the firman naming General Zahedi as prime minister, statements of the Shah at Baghdad, and an interview from the hidden Zahedi that he was the legal prime minister and that the so-called *coup d'état* was a device thought up by Mossadeq to divert public opinion. Members of the government and the National Movement met during the day to discuss the formation of a regency council, while within the city street names were changed from Reza Shah to Republic and from Pahlavi to People. The Shah flew on to Rome. However, during that afternoon appeared the first pro-Shah demonstrations, while both nationalist elements and the armed forces pitched in to break up

Tudeh demonstrations—apparently the nationalists were beginning to feel that the communists would be the chief gainers from excesses and disorder.

Early on the morning of August 19 a ground swell began to roll up from south Tehran—the older section of the town—as a number of groups armed with crude weapons converged upon the official and business center of the capital. Confronted by armed forces, their shouts of "Long live the Shah" won over soldiers and police. Tanks joined the swelling columns, as did unarmed soldiers from the barracks. By noon the offices of all the newspapers which attacked the monarchy had been ransacked as had the headquarters of the groups supporting Dr. Mossadeq. Other groups of mixed civilian and military members stormed the ministerial buildings. A long column wound its way north of the town to take over Radio Tehran and at two in the afternoon this station announced that General Zahedi headed a new government. At this same time the major fighting of the day broke out at Mossadeq's residence, but it was not until near the end of the afternoon that he once more fled over the garden wall. Before dark Zahedi had been heard over Radio Tehran and quiet reigned. Broadcasts from the provincial stations offered loyalty to the Shah and the new government. Within the next few days Dr. Mossadeq, members of his former cabinet, officials, editors and many members of the Tudeh party were placed under arrest, but Husayn Fatemi remained in hiding—aided by the Tudeh party—for seven months.

On August 22 the Shah returned to a tumultuous welcome and spoke of the necessity for the government to concentrate all its attention upon social reforms. By this time it was perfectly clear that a spontaneous uprising of the masses, including thousands who were not normally involved in parties or politics, had occurred. The Tudeh party, possibly acting on instructions from its Soviet masters, made a serious tactical error in pushing for revolution when it lacked the power base to carry it out. The insults of Fatemi, the destruction of the statues of the ruler and his father, and the shouts in favor of a republic had an effect quite opposite from that intended for they had aroused fears as to the future of the country and the wisdom of casting aside long established, national traditions. In actual fact, the people of Tehran—and other cities as well—had been forced to make a choice between the stable institution of the monarchy and an alternative which seemed to offer instability and possible political chaos. Statements that the American and British intelligence services played active roles in support of the monarchy in these critical days have won acceptance, although the Shah, in his own published record of these days, gives all the credit to his loyal followers.

General Zahedi held the office of prime minister until April 1955, heading a government that was unspectacular in its operations, but more stable than any of prior years. Funds flowed to Iran from the United States to support economic stabilization after the near bankruptcy of the Mossadeq period, and to aid in economic development plans. Early in 1954 the Anglo-Iranian Oil Company announced that it was meeting with seven other large oil companies with interests in the Middle East to discuss ways of resolving the difficulties which prevented Iranian oil from returning to the world markets and about this same time the American government stated that should five American oil firms join a consortium they would not be liable for anti-trust prosecution. Technical experts representing the eight companies flew out to survey the condition of the fields and refineries in Iran.

Elections for the XVIIIth Majlis and the IInd Senate began in January and were conducted with far greater expedition than usual, for by mid-March a sufficient number of deputies had been chosen so that the Majlis could open with the required quorum. All of these deputies had asserted their loyalty to the Shah, and their number included no known supporters of the National Front. On April 10 the oil consortium was formally established at London and representatives left for Iran. Negotiations went forward, stalled on complex issues, and then were resumed over the next months. The major problem was that of how any agreement could be fitted within the rather rigid terms of the Nationalization Law. In the end, persistence and good will won out and on August 5 the Iranian delegation and the International Consortium announced full agreement in principle, subject to putting the agreement into legal form and having it ratified by all parties concerned. The companies making up the consortium were the Anglo-Iranian Oil Company (which name was changed to The British Petroleum Company Ltd. in December 1954), Gulf Oil Corporation, Socony-Vacuum Company, Standard Oil of California, Standard Oil of New Jersey, The Texas Company, The French Petrol Company, and Royal Dutch Shell, and it was their belief that in the first three full years of operation under this agreement Iran would derive a total direct income of £150,000,000. The legal form of the agreement, nearly 50 pages long and extremely technical throughout its many articles, was presented to the Majlis for ratification on September 21 by General Zahedi and Dr. 'Ali Amini, Minister of Finance and chief oil negotiator for Iran. After full discussion, on October 21 some 113 deputies voted in favor with 5 opposing the agreement and soon thereafter it was also ratified by the Senate. Tankers began to load at Abadan at once, while on November 2 the United States announced that it would put the sum of $127,300,000 at the disposal of the Iranian government: $85,000,000 as a loan and the rest as free assistance, the purpose being to permit Iran to pursue its development program pending adequate receipts from oil revenues.

Throughout the year the government—employing the means common to recent years of continuing martial law and strict control over the press—met with little opposition. Elements of the Iran party, the Third Force, and the Pan-Iranists regrouped into a rather ineffective "National Resistance Movement," dedicated to the return of Dr. Mossadeq to power—he had been tried by a military court on charges which included failing to obey imperial decrees and plotting to overthrow the constitutional monarchy, found guilty, and sentenced to three years in prison. While Zahedi had declared himself to be the executor of the national movement, he had not won over the intellectual and middle class supporters of Dr. Mossadeq. Elements of the faculty and student body at Tehran University and other institutions also continued faithful to Mossadeq as did many Iranians pursuing their studies abroad, who described the recent regime as the "republican" period in Iran. The Tudeh party struggled to analyze its own errors, to regroup its dependable elements, and to try to attract all opponents of the government into a popular front: the larger subject of the policy and activity of the party is discussed in a following chapter. On September 20 military authorities at Tehran made the first detailed announcement of the discovery of an extensive penetration of the armed forces by the Tudeh party and hundreds of officers and men were arrested. A flood of confessions provided material for establishing the five different sections of the party within the armed forces as well as defining their activities which ranged from advocating the overthrow of the constitutional monarchy, to sabotage, and to training for partisan warfare against the government. Those arrested were tried in groups and more than a score of officers were executed. As an immediate reaction the government proposed a bill carrying stiff penalties for members of organizations which were collectivist in nature, which were against the Muslim religion, or which attacked the constitutional monarchy.

Husayn 'Ala succeeded General Zahedi as prime minister in April 1955 and remained in office until April 1957: he turned his attention to domestic reforms which had not been carried out by his predecessor. On October 11 Iran adhered to the Baghdad Pact, initiated in February 1955 by Turkey, Iraq, Pakistan, and the United Kingdom. November 17 Prime Minister 'Ala was slightly injured by a would-be assassin; on this occasion the government took rapid and stern action.

In March 1956, at the beginning of the Persian year, the Shah pardoned a number of former Tudeh party members and former collaborators of Dr. Mossadeq. At the end of March the Development and Resources Corporation of New York signed a five-year agreement with the Plan Organization covering the agricultural and economic development of the southwestern region of Khuzistan. In May the Shah made a state visit to Turkey, and after his return opened the XIXth Majlis, speaking of the necessity of strengthening the armed forces, making rapid progress in economic development, and putting the oil revenues to the most effective use. In June the

Shah and Queen Soraya paid a state visit to Moscow: the ruler, in conversations with the Soviet leaders, defended Iran's membership in defensive alliances. This visit was followed by a rather dramatic change in the Soviet attitude toward Iran: official criticism of the ruler and of the successive governments almost ceased and the USSR expressed friendship and its desire for closer economic relations.

Early in January 1957 Iran subscribed to the Eisenhower Doctrine with official enthusiasm. In the same month the publication of a lengthy report of hearings before a committee of the House of Representatives brought to light severe criticism of the manner in which American economic aid had been dispensed in Iran: in rebuttal the Department of State asserted that this aid had stabilized the financial and political structures of the country. In April the Shah named Dr. Manuchehr Eqbal as prime minister. While Eqbal was to remain in the post for a record 1,245 days, his appointment marked a period in which the ruler took charge of the operation of the government—Eqbal did not hesitate to inform the Majlis that he was the servant of the Shah. In the political arena the ruler sponsored the formation of two new political parties, the *Melliyun*, or Nationalist, Party, and the *Mardum*, or People's, Party. The first of these was headed by Dr. Eqbal, and the second by Asadullah Alam, a long-time associate of the ruler. The Shah stated that the government must be supported by a majority party in the Majlis, which was the Melliyun Party, while the Mardum Party was to fulfill the functions of a loyal opposition, free to criticize the internal policies of the government but not the conduct of foreign affairs. The ruler was aware of the difficulty of creating parties of principles by fiat but insisted that a start had to be made and that in time each group would attract an active following and gradually assume the role of a truly democratic and responsible party. Sessions of the Baghdad Pact nations were held at Karachi in June. The United States, previously a member of the Economic Committee, joined the Military Committee, although it continued to refrain from taking up full membership: in the meetings emphasis was placed upon the construction of telecommunications and highway and rail links between the member countries. Later in the year the Parliament passed a bill providing for the establishment of a National Security Organization—known as Savak after the initials of its name in Persian—which was to centralize the control of political subversion and of counterespionage activity. It was headed by General Timur Bakhtiar.

In March 1958 the Shah, responsive to the advice of a council of elder statesmen, wrote to childless Queen Soraya, then in Europe, and obtained her consent to a divorce: they had been married in 1951.

In April the ruler established the Bonyad-i-Pahlavi, or the Pahlavi Foundation. Resources of the crown, other than agricultural lands, were turned over to the foundation, and the income was to be expended for public health, education, and social services. These resources included a number of hotels, shares in the ownership of local industries, and the National Insurance Company. The revolution of July in Iraq had an immediate impact upon the government of Iran. The Shah began to advocate an over-all program of land reform, stating that the regime in Iraq had been overthrown because it failed to undertake the distribution of land. In addition, he set up an Imperial Investigation Commission which was to receive complaints from the public against public officials and to see that proper action was taken. This general concern carried over into the opening months of 1959 when the Majlis passed two bills designed to provide the material required for the prosecution of appointed and elected officials engaged in illegal or corrupt practices; such officials were required to submit statements of personal and family assets. The second of these was soon known as the "Where did you get it?" law: within a year well over 2,000 officials had been brought before courts and then dismissed from government service.

Early in the year Iraq withdrew from the Baghdad Pact, and the remaining members announced that its name had been changed to that of the Central Treaty Organization (CENTO). On March 5 Iran and the United States signed a bilateral agreement that provided for American aid in case of aggression against Iran: this agreement was bitterly attacked by the Soviet Union. On December 14 President

Eisenhower visited Tehran and praised the contributions of Iran to regional stability. Finally, on December 21 the Shah was married to 21-year-old Farah Diba.

The event of 1960 which drew the most attention was the elections for the XXth Majlis, held in August after weeks of hectic campaigning in which Dr. 'Ali Amini came to the fore as spokesman for the independent candidates. When the results were announced, some two-thirds of the new deputies belonged to the Melliyun party and the balance, with the exception of a few independents, to the Mardum party. However, heated charges of a pre-election agreement between these parties and of discrimination against independent candidates led the Shah to request the deputies to resign. The heads of the parties also resigned their posts, with Dr. Eqbal also resigning as prime minister to be succeeded by Jafar Sharif-Emami. On October 31 Queen Farah gave birth to a boy, Reza Cyrus 'Ali: the nation rejoiced, and the ruler was delighted to have finally produced an heir to the throne.

In January 1961 a second round of elections was held amidst charges of rigged elections and featuring a student demonstration that resulted in the closing of Tehran University. The ruler opened the XXth Majlis on February 21: it included 70 members of the Melliyun party, 60 from the Mardum party, and about 20 independents. The National Front which had boycotted the elections continued to attack the government, and on May 4 backed a massive demonstration in support of striking teachers. Apparently disturbed over a lack of firmness in handling the opposition and the absence of a progressive policy, on May 5 the Shah dismissed Sharif-Emami and named Dr. 'Ali Amini as prime minister. On May 9 the ruler dissolved the Majlis. In selecting Dr. Amini, the ruler had picked a man who would not be his "servant," but who would form his own policies and stand or fall by his success or failure in carrying them out. His announced program of reforms placed special emphasis upon the elimination of corruption, the distribution of land to the peasants, and upon efforts toward financial stabilization.

The National Front insisted that the article of the constitution which provides that new elections must follow shortly after a Majlis has been dissolved must be enforced, and made plans to hold a massive demonstration at Tehran on July 21. The previous day many of the leaders of the National Front were arrested, while those who gathered at the site of the meeting were dispersed by the police.

On January 15, 1962, the Shah issued a decree entitled "Agricultural Reforms." Replacing a law passed by the Parliament in May 1960 which limited the acreage which could be held by an individual, the decree stated that no person could own more than one agricultural village, and provided that excess holdings would be purchased by the state and then resold to the peasants living in the villages. Implementation of the program of land distribution was assigned to the Minister of Agriculture, Dr. Hasan Arsanjani, and in March the Shah distributed deeds to land to peasants in Azerbaijan. Throughout the year this program gained momentum.

On July 18 Prime Minister Amini resigned, blaming tardy American economic aid and the reduction of military assistance for his action. The Shah named Asadullah Alam to take his place.

In January 1963 over six million voted in a referendum, approving by about 12 to 1, six reform measures sponsored by the ruler. These were, the Land Reform bill, the sale of state-owned factories to finance land reform, sharing of workers in up to 20 percent of industrial profits, nationalization of forests, amendment of the election law to include female suffrage, and establishment of a Literacy Corps. These measures were followed in subsequent years by a Health Corps, a Reconstruction and Development Corps, Houses of Justice, nationalization of water resources, and the decentralization of administration. In its entirety these actions were described by the regime as the Shah's White Revolution, or the Shah-People Revolution.

In October the Shah opened the XXIst Majlis: the country had been without a parliament for 18 months. In December Hasan Ali Mansur presided over the founding meeting of the *Iran Novin* (New Iran) Party. Replacing the Melliyun Party, it pledged to support the Shah-People Revolution and its members soon included a majority of the Majlis and many high officials. On March 7, 1964, Hasan Ali Mansur became prime minister, charged to push forward the reform program. On January 21, 1965, Prime Minister Mansur was shot by a student, alleged to be a member of the Devotees of Islam. When he died on the 26th, his place was taken by Amir Abbas Hoveida, a cabinet minister and second in command of the Iran Novin Party. Early in April a conscript in the royal guard attempted to assassinate the Shah. Although he was killed by other guards, a court found that individuals allied with the Tudeh Party had planned the attempt. In October an agreement was concluded whereby the USSR would construct a steel mill and a machine tool plant in Iran and would receive natural gas by pipeline from the southern oil fields of Iran. Near the end of the year, and again in 1966, it was announced that Iran was making heavy purchases of wheat from abroad.

In September 1967 a constituent assembly designated Empress Farah, in Iran called Shahbanu Farah, as regent, to act in the event of the death of the Shah until Crown Prince Reza became twenty years of age. On October 6 the sessions of the XXIInd Majlis and the Vth Senate were inaugurated, and on the 26th the long postponed coronation of Muhammad Reza Pahlavi took place. Following the example of his father's coronation, he placed the Pahlavi crown on his head, while Crown Prince Reza, almost seven, stood beside the throne. Then he crowned the Shahbanu. In November the United States program of aid to Iran which had totalled over $600 million was terminated by mutual consent in view of the increasing economic prosperity of the country, and at the end of the year the Majlis authorized a special fund of $260 million to strengthen the armed forces, especially in the region of the Persian Gulf. Emphasis on the armed forces continued: in January 1968 the Shah and the Shahbanou were in the United States, with the ruler receiving assurances from President Johnson that the United States would support the building of an adequate defense force. In October 1969 the Shah was again on an official visit to the United States, and conferred with President Nixon. In fact, the ties of Iran to the United States were less strong than in earlier years as the ruler responded to overtures from the USSR, bought military equipment from other countries, and broadened Iran's political and economic relations with other nations, including the People's Republic of China.

In the most recent years internal politics and the march of events are of less importance than a number of enduring problems and issues which absorbed much of the attention of the Shah and the government. However, a few highlights may be mentioned. In September 1971 the XXIIIrd Majlis and the VIth Senate were convened, and in October the 2,500th anniversary of the establishment of the monarchy in Iran was magnificently celebrated at Pasargadae, Persepolis, and Tehran at a cost of nearly $17 million. An elegant tent city at Persepolis housed the heads of states of many nations. Along with other guests, they witnessed a stirring historical pageant: units of the Iranian forces of earlier periods, clad in the costumes and uniforms of those times, marched past the massive remains of the site. At Tehran the ruler dedicated a new 100,000-seat stadium, and the Shahyad monument, a towering structure which dominates an approach to the capital.

On May 30, 1972, President and Mrs. Nixon arrived in Tehran from Moscow for a visit of two days. In the official communiqué following talks between the President and the Shah it was stated that both sides believed that the security and stability of the Persian Gulf were of vital importance to the littoral states and that these states were mainly responsible for preserving this security, and the President confirmed that the United States would continue its cooperation with Iran to strengthen its defensive capabilities.

In January 1973 Prime Minister Hoveida began his eighth year in that post—an all-time record. Plagued, as are many other countries of the world, with a rising traffic in narcotics, in 1969 the government lifted a ban imposed in 1956 on growing opium poppies to permit limited crops under strict controls, and, at the same time, imposed the death penalty for the smuggling of narcotics. In the months which followed many smugglers were arrested, quickly convicted, and executed.

Terrorism and sabotage struck the country. SAVAK, the National Security Organization, identified the directing groups; the Iran Liberation Organization, the Revolutionary Organization of the Tudeh Party, and the Revolutionary Branch of Iranian Communists, with the latter known as SAKA. Their members engaged in hijacking planes, robbing banks, setting off bombs, attacking police stations, attempting kidnappings, and assassinations. Many of the terrorists were caught, tried, and convicted and some twenty-five were executed; in 1973 some remained active. Some members of the groups were given guerrilla training in Iraq and elsewhere, while others were alleged to have ties with dissident Iranian students abroad. In January 1971 the government announced that the Confederation of Iranian Students Abroad, active in several European countries and the United States, was considered to be a communist organization, and called on its members to resign or to be held legally responsible for their continued affiliation.

The economic scene was particularly bright. Revenues from oil soared as the consortium and other concessionaires stepped up production, and as negotiations increased the base figures upon which the returns to Iran were calculated. These funds, together with loans from abroad, were channeled into the economic infrastructure with the construction of major industrial complexes and scores of manufacturing plants. From 1967 to 1974 the national income grew by between 10 percent and 14 percent annually—a remarkably high rate—while the cost of living rose very slightly. Only the production of wheat and of livestock for meat was unsatisfactory.

In the field of foreign relations Iran was engrossed with the region of the Persian Gulf. On the grounds of its much earlier sovereignty, in the 1930s Iran laid claim to the Bahrain islands, and when Dr. Mossadeq was prime minister threatened to send an expeditionary force to recover the islands, then under special treaty relations with Great Britain. Then, in 1957 the islands were declared to comprise the fourteenth province of the empire of Iran. A few years later the rising tide of Arab nationalism was marked by the action of Egypt and Iraq to rename the Persian Gulf the Arab Gulf, and by their propaganda efforts to separate the province of Khuzistan from Iran, as well as by the activity of subversive organizations sponsored by Iraq.

In January 1968 the British government announced that it would withdraw all its military forces from the Persian Gulf by the end of 1971—forces that were based at Bahrain and in the Arab amirates. The Shah then stated that Iran was the state best fitted to exercise leadership in this region, that no other great power should attempt to fill whatever vacuum was created by the British withdrawal, and that Iran was prepared to enter into a regional pact or treaty with other littoral states. Realizing that Iran's future role in the Gulf would require the good will of all the Arab states and amirates, the Shah stated that his country would not use force to press the claim to Bahrain, and that the people of Bahrain should decide their own political future. In March and April 1970 a fact-finding mission from the United Nations determined that the people of Bahrain wanted an independent state, and the country's declaration of this independence in August 1971 was immediately followed by its recognition by Iran.

Perennial difficulties with the regimes of Iraq escalated in April 1969 over navigation rights in the Shatt al-Arab, the confluence of the Tigris and Euphrates rivers of Iraq and the Karun river of Iran, and the regime of Iraq reacted to a firm stand taken by Iran by expelling a very large number of Persians long resident in that country. Happier relations were maintained with the other Arab states, and especially with Saudi Arabia. These relations were marked by agreements between Iran and several of these states on the demarcation of the median line of the Gulf as the boundary between contiguous or facing states. Iran insisted on its ownership of the tiny islands of Abu Musa, Greater Tunb, and Lesser Tunb at the entrance to the Gulf as against the claims of two of the amirates, and on November 30, 1971, armed detachments occupied these islands. Iraq reacted by severing diplomatic relations with Iran. Currently Iran is engaged in an impressive program of constructing ports and other facilities along its Gulf coast, as well as in strengthening its naval and air forces in the region.

V. THE PEOPLE AND
THEIR LIFE

Population and Ethnic Elements

DRAWING on the accounts of earlier travelers, it has been estimated that during the Safavid period the population of Iran was about 40 million. Wars, famines, and epidemics are thought to have reduced this number to 10 million by the middle of the nineteenth century. The reality of the economic renaissance of the country is shown by the facts that the census of 1956 reported a population of 18,954,000, and that of 1966 25,781,000. In 1978 the population was reliably estimated to number 35,500,000.

In 1956 29 percent of the people lived in towns, while in 1978 this figure had attained 48.7 percent. Thus, in 1978 the rural population was 51.3 percent of the total and included farmers and craftsmen living in 60,000 villages and the migratory and semi-nomadic tribes. The migration from the villages to the cities continues to grow as families move to urban areas in search of employment, and to enjoy the amenities not to be found in the villages. While Tehran has attracted the largest number of migrants, the current decentralization of industry brings many to other cities. Tehran has over 4 million inhabitants. Isfahan has 520,000; Mashhad 510,000; Tabriz 465,000; Shiraz 325,000; Abadan 300,000; Ahwaz 260,000; and Kermanshah 225,000. Several other cities have about 100,000 people.

The average density of population is 47 to the square mile, very close to that of the United States. Throughout the great deserts and much of the southeastern section of the country the average is less than 10 to the square mile, while in the northwest and along the Caspian coast the figure is over 100.

The current population increase is 3.2 percent a year. The average age of marriage for men is 26 years and for women nearly 19 years: a special legal permit is required for marriages of girls under 15 and men under 18 years of age. In both urban and rural areas the average family has five members. Polygamy, sanctioned by Islam, is on the wane as the Family Protection Law requires judicial consent for pluralistic marriages. The same law deprives men of the former privilege of unilateral divorce. The divorce rate remains rather high, one divorce to every ten marriages, and would be higher if it were not for marriage contracts which provide for the return of the bride-price to the divorced wife.

As is now common in many countries where improved health measures have brought about a decline in death rates and there is a high rate of population increase, young people make up a very large proportion of the population: in Iran 55 percent of the population is under twenty years of age, with 25 percent below the age of twelve. More than 1,500 Family Planning and Health Centers are active throughout the country in an intensive program of population control.

As has been noted earlier, the "Iranian" population of the country has been augmented by other ethnic elements. Arabs who entered the country during and after the seventh century A.D. are now found in the province of Khuzistan and along the Persian Gulf littoral. From the tenth to the fourteenth centuries Turkish tribes moved into the northern areas of the plateau in considerable numbers. The inhabitants of the Caspian Sea littoral were long isolated from the plateau by the towering Alborz range, and still cling to distinctive languages which stem from earlier forms of Persian. The numerous nomadic tribes further complicate the ethnic and linguistic structures of the country. Some of them, judging from the anthropological evidence of their head types, may have been resident in Iran for many centuries, others moved into the region in later times, and still others have moved or have been transferred from one part of the country to another. Some of the tribes speak dialects which derive from older forms of Persian, while others speak languages which are not consistent with their ethnic origins.

About two-thirds of the entire population speaks Persian, in many cases as a second language. One-fifth of the total population, dwelling in the densely-settled northwest province of Azerbaijan and thence southward in a large triangle reaching almost to Tehran, speaks Turki, which has a simple grammar and has absorbed many Persian words. A goodly number of these people acquired the language following the Mongol conquest or in more recent times. While it is hard to draw the line between the Turkish elements and the Iranian elements who now speak Turki, it is certain that the Turkish element is composed of Turanian Turks who derive originally from central Asia and have little racial affinity with the Ottoman Turks of the present Turkish Republic. The Qashqa'i tribe of the southwest speaks a dialect very similar to Turki, suggesting that this group migrated from the northwestern corner of the country.

The population of Iran therefore reflects the mass movements into the region and the residue from recurrent invasions, but at the same time there is rather more homogeneity of population than in most other countries of comparable size. There is also a basis for national unity and national self-consciousness in the overwhelming adherence of the community to the Shi'a sect of Islam and in the all-pervading influence of Persian literature, culture, manners and customs, and way of life.

Ninety-eight percent of the inhabitants are Muslims, and 90 percent of this number are Shi'as. The Sunni sect has many followers among the Kurds, Baluchi, and Turkoman. There are relatively few minority religious groups within the country, and the Iranians have always been tolerant and shown comparatively little of the discrimination of other lands toward minor racial or religious elements. In the northwest and at Tehran and Isfahan are some 250,000 Armenians, who are also prosperous merchants in the large towns, and some 25,000 Nestorian Christians. Some 80,000 Jews are settled in the large towns and in certain ancient farming communities; many members of this community have gone to Israel in recent years. There are about 50,000 Protestants and Roman Catholics. An estimate places the Bahais at 60,000. At Yazd and Kerman, and in fewer number at Tehran, Isfahan, and Shiraz are colonies of Parsis, Iranians who still hold to the religion of Zoroaster and still worship in fire temples. The Parsis, or Zoroastrians, are famed as gardeners and merchants. They number about 30,000.

The Persians are brunets, with dark brown or black hair and dark brown eyes, and of medium height and build. Complexions range from quite fair to swarthy, with general coloring similar to that of the Italians or the Greeks.

The Nomadic Tribes

The census of 1966 gave the total of these tribes as 516,583, a figure of very questionable veracity, since estimates by specialists place their number at two million or more. Generally speaking, these groups ring the periphery of the country, with the exception of the littoral of the Caspian Sea. None of them have shown any desire to detach themselves from Iran, in spite of subversive propaganda by the Soviets among the Kurds and by other elements among the Arabs and the Baluchi. They may be named beginning in the northwest and following along the periphery. Within the general region of the Zagros range which runs to the south are the Shahseven, Kurds, Lurs, Bakhtiari, Qashqa'i, Mamassani, and Kuh Galui. Arab tribes are found in Khuzistan and form part of the Khamseh confederacy. In southeastern Iran are the Baluchi, and to their north the Brahui. There are other Kurds in Khorasan who were moved there over a century ago to help protect the area from external pressures, and the plains to the east of the Caspian Sea are occupied by several Turkoman tribes, noted for their splendid rugs.

The Kurds, Lurs, Bakhtiari, Kuh Galui, Mamassani, and Baluchi speak dialects which may be related to Old Persian or to other ancient Indo-Iranian languages; the Arabs speak Arabic; the Shahsevens, Qashqa'i, and a part of the Khamseh group speak Turki; the Brahui speak a Dravidian tongue; and the Turkoman their own dialect of Turkish. It has been suggested that the Qashqa'i and the Turki-speaking people of the northwest acquired this language in place of their former Indo-Iranian tongues.

These nomads and many lesser groups are the most picturesque element of the Iranian population, representing the best physical type with a bravery which has always furnished the stiff backbone of Iranian armies, and several ruling dynasties have sprung from tribal warriors. A tribe, called *eel* after the Turkish term, may be very numerous or may number only a few hundred. In the case of the Qashqa'i, who may total some 150,000, the tribe is divided into numerous smaller divisions or parts, with a division either a unit or itself subdivided into ten, twenty, or fifty smaller groups. Such subtribes or clan groupings are variously known as *tayifeh*, *tireh*, and *dasteh*. In several regions the tribal leader is known as the *khan* and the heads of the sub-tribes as *kalantars*. Certain tribes claim ownership of large areas, basing their claims on awards of *tiyul*, or fief, made by earlier rulers of Iran, frequently in return for military service.

According to season the tribes move from their *garmsir*, or winter quarters, to the *sardsir*, or summer quarters. These Persian terms are paralleled by the Turkish terms, *qishlaq*, and *yeylaq*. In the fall the tribes of the Zagros range move down to the warm low plains along the Iraq frontier and toward the shores of the Persian Gulf to sow their cereal crops. In the spring, leaving behind some members of the tribe to reap the harvest, they migrate again into the highest mountain valleys where their flocks can find good grazing throughout the summer. The seasonal migrations may cover distances of more than two hundred miles and may take weeks. They follow an *eel-rah*, the tribal road, which winds its way alongside stretches of cultivated land, and sources of water for the flocks. The rate of movement is dictated by the slow progress of the flocks of sheep and goats, which the men and boys drive ahead while the women and children ride perched in precarious fashion atop the baggage lashed to donkeys and camels. Each evening the tents are pitched along the side of the road. Visitors to Iran may expect to see the tribes only at the times of their migrations, for their summer and winter quarters are usually remote from the frequented highways.

Family life is pursued in the black tents which are made of a tough goat's-hair cloth woven by the women of the tribes. Their vertical side-walls and slightly sloping tops are supported by stout poles, and the tent can be quickly struck and packed in a small space for transportation from one site to another. The furnishings are extremely simple. The floor coverings are usually rugs woven by the women, or thick felt mats; the blankets are piled against one wall and along the other walls are placed the copper utensils, goat-skin containers for liquids, earthenware jars, bags of grain, and occasionally a wooden chest for clothing.

The flocks themselves sustain the nomadic life. They furnish milk, butter, and cheese, and their wool is used for tribal weaving. The tribal lands occupied in the summer often contain orchards and groves of nut-bearing trees which are an added source of cash income. Scrub oaks are burned to obtain charcoal. From time to time families head for the nearest town to trade clarified butter, wool, lambskins, rugs, and animals for flour, sugar, tea, dates, fruits, piece goods, clothes, shoes, and transistor radios.

The fascinating jewelry worn by the women usually include strings of old silver and gold coins, and represents the whole accumulated wealth of the family group. Other valuable property of the tribe is the herds of horses and the guns; the men are fine marksmen and very proud of their weapons.

The men do the hunting and care for the flocks and the herds of horses, and the women gather fuel, carry water, and do the cooking, sewing, and weaving. Polygamy is comparatively rare among the tribes, and the women are not veiled. The social life of the tribe centers around the guest tent of the tribal chief. There the tribal dances are held to the stirring rhythm of flute and drum, ballads are recited, stories told, and news, gossip, and rumor exchanged. Acquaintances and strangers are welcomed to the generous hospitality of the guest tent. Tribal boundaries are fairly well defined, and in recent years intertribal raiding and warfare have been on the decline.

Much still remains to be learned about the tribes. Until fairly recently interest was primarily in their military potentialities, with attention centering on determining the exact names, numbers of tents, and amount of arms belonging to each tribe, sub-tribe, and clan. However, in very recent years trained anthropologists have lived with several of the tribes and published accounts of tribal organization, customs, and modes of life, while a member of the Qashqa'i has produced a detailed study with emphasis on tribal codes of conduct and of justice. Collections of their songs and stories have been published at Tehran. The Ethnographical Museum at Tehran has several rooms devoted to tribal life and culture, including fully equipped tents and life-sized figures of men and women clad in everyday costumes and in the rich and brilliant cloth-thing worn at weddings and on festival days.

The nomadic tribes and the settled population were long mutually distrustful, if not actively hostile. Rulers sought to enlist tribal warriors as mercenaries in times of internal disorder or of foreign threats, and then when they became too strong transferred entire groups to regions remote from their habitats. During the reign of Reza Shah efforts to establish the authority of the government throughout the country gave rise to a number of tribal revolts which were severely repressed. After 1930 attempts were made to collect their arms, and to settle groups on the land by moving them into sites where housing had been provided. As soon as circumstances permitted, they went back to their traditional ways of life and the settlements crumbled into ruins.

The policy of successive recent governments has been to encourage the settlement of the tribes. Difficulties persist. On the one hand, good agricultural land is not available for their numbers, and, on the other, the tribes despise farming, and must be persuaded that the material benefits of settled life would outweigh its disadvantages. However, increased settlement does occur as newly irrigated lands are made available to nomads in Azerbaijan and Fars. Already, the fact that these and many other areas have been taken out of the lands used for pasturage has resulted in a decline in the production of meat.

The freedom of action of the tribes has been curtailed. Army officers share authority with the traditional tribal leaders, licenses are required for tribesmen to bear arms, and the gendarmerie controls migrations.

Villages and Village Life

The countless farming communities of Iran exist wherever water for drinking purposes and for irrigation is available, and where the soil is suitable for growing crops. In those regions where adequate sources of water supply are far apart the villages are concentrated near them and therefore tend to be larger, while in sections where mountain streams, springs, and qanat lines are available small hamlets dot the landscape at intervals of only two or three miles. The traveler along the main roads will see many such villages, but by far the largest number are hidden away in remote mountain valleys. He will also see ruined and deserted villages, evidence of the economic decline which has been arrested only in the last decades.

Only along the shores of the Caspian are isolated houses set in the midst of the fields; in all other sections of the country the farmers live in compact and crowded settlements and work in fields which may lie at a considerable distance from the village. Until a few years ago all the villages were protected against marauding bands by high walls of mud brick, pierced by a single bastion-flanked doorway. Today such walls are no longer necessary, and only occasionally are they still standing in good condition. Within each village is a haphazard network of narrow lanes, sometimes paved with cobblestones, and usually one straight main street along which runs the water channel which is the sine qua non of village life. There is usually a village square, an irregular open space made conspicuous by a few towering trees and fronting on it the village mosque or the domed tomb of some local saint. At one edge of the village is always the community threshing floor.

The narrow lanes are bordered by high mud walls, in which double doors lead into the courtyard of each house. The typical courtyard contains a few trees and flowers, and a pool, filled at intervals from the main channel passing through the village, which provides water for all domestic purposes except drinking. The houses are built of mud brick, the type of construction varying with each region. In areas where timber is plentiful crossbeams and overlapping straw mats support a flat roof of earth, which after each rain is packed down with a stone roller kept on the roof for just that purpose. The flat roofs provide a cool place for the family to sleep during the hot summer months. In other sections of the country the houses are roofed with tunnel-vaults or domes in mud brick which are erected by local masons, following an age-old technique, without any supporting scaffold.

The houses usually occupy one short side of the court and contain three or four rooms, but larger dwellings occupy two or three sides. The houses face toward the south so that the warmth of the winter sun can carry directly into the main rooms, and a porch across the front of the house intercepts the rays of the high summer sun so that the rooms are shady all day during the hot months. The better houses have one reception or living room in which the cherished possessions of the family are on display. Rugs, often locally woven, cover the floor, and quilts, mattresses, and pillows are neatly piled in a niche in the wall. An oil lamp stands in another niche. There is always a mirror on the wall, and usually a few bright lithographs of religious subjects or colored pictures from Persian or foreign magazines. There are no chairs or tables or family heirlooms. At mealtime a cloth is spread on the floor and the men of the family gather around food served in dishes on a large brass tray, and at bedtime the mattresses and blankets are spread out and everyone retires early.

The large houses have a vaulted room below the main part of the building which remains pleasantly cool even in the middle of the summer. In the courtyard is a brick oven where the characteristic flat sheets of bread are baked by the women of the house, and where the family cooking is done in copper vessels over a charcoal brazier. At one corner of the courtyard is the simple sanitation arrangement which is much like the outhouse of any country. If the family owns a pair of oxen and a donkey these animals are kept in a separate section of the courtyard, although in some regions the entire ground floor is given over to the farm animals and the living quarters of the family are all on the second floor.

Many villages have a bath. The building has a series of rooms which are mostly underground so that only the upper part of the domes with their glass lighting apertures project above the surface. One man is put in charge of the community bath and it is his duty to keep it clean and see that steaming hot water is available for a given number of hours each week. The villagers can bathe as often as they wish, and everyone helps to support the bath attendant by contributions of wheat, straw, fuel, and fruit.

Usually each village has a mill in which a water-wheel turns one great millstone against another. The miller grinds wheat and barley for the community and receives from each villager a share of the flour ground.

Each village has a *kadkhuda*, or headman. In very early times he was elected by the villagers, later named by the landlord, and after 1937 appointed by the local governor. He may be aided by a group of village elders, the *rish safid*, or white beards. A *dashtban*, or watchman, keeps an eye out for thieves and wild beasts, and the *mirab* is in charge of supervising the network of water channels, diverting the proper amounts at established intervals to houses and fields. Intravillage and inter-village violence flares most frequently from disputes over the distribution of water.

Shops are less common than baths and mills. When they do occur, their stock is largely limited to sugar, tea, tobacco, rice, spices, needles and thread, nails, salt, dyes, cotton piece goods, matches, kerosene, lamp chimneys, and—increasingly—parts for irrigation pumps and radios. Travelling pedlars move among the smaller villages. The number of village schools is on the rapid increase. Funds for construction of buildings and for teachers' salaries are now made available.

On the outskirts of the large villages a few houses stand among orchard or groves of poplar trees, and then the farming land begins abruptly just beyond the last walls of the village. The villagers keep their own sheep and goats and in the summer make up a village flock which is sent to the mountains in charge of the young men of the village. Most of the sheep are of the fat-tailed variety which look so comical when they are seen for the first time. An early traveler to Iran has a story about them which seems quite worthy of belief: "The Sheep are prodigiously large, trailing Tails after them of the Weight of Thirty Pounds, full of fat, which sometimes prove such Incumbrances that unless small Carts with two Wheels were provided for their Carriage they would trail upon the Ground and wound themselves against every sharp Stone and rough Place of the Ground."

The staple food of the village people is bread or rice, according to the section of the country, *mast*, our yoghurt, cheese, and clarified butter. Eggs, chickens, a very small amount of mutton, onions, cucumbers, radishes, melons, fruit, nuts, and tea round out their simple diet. Mast churned with butter makes a very popular drink called *dugh*. Special events call for the more elaborate foods which are part of the regular diet of the town dwellers.

The farmers may wear a modified version of European dress, but the native costume of cotton shirt, baggy black or blue cotton trousers and a long blue cotton coat is generally worn. The women wear black trousers which are gathered at the ankles, a shirt, and a length of cotton piece goods draped around the body and over the head so that it serves as both head covering and veil, although the village women were never closely veiled as were the women of the towns. There is a natural division of labor; the men till the fields and the women look after the house and weave rugs. Marriages are arranged between heads of families, and the wedding celebrations are among the gayest occasions in village life. The groom displays his possessions in cash and kind, and the bride's family supplies quilts, clothing, and simple housekeeping equipment.

The inherent virtues and qualities of these village people can scarcely be too highly recommended. They have a quick natural intelligence, a ready sense of humor, and a lively interest in the world about them. They are extremely hospitable and friendly, and will place all their meager resources at the disposal of their honored guests. However, they view the townspeople with considerable distrust, possibly because they were long exploited by officials from the towns, and taken advantage of in their trips to the bazaars of these towns, and have them categorized with epithets which are often uncomplimentary. Thus, the people of Kashan are said to be cowardly and complaining, those of Isfahan shrewd and grasping, those of Shiraz more enlightened than most because of its good climate, those of Mashhad tricky and untruthful, those of Semnan frugal, and those of Tabriz aggressive and brave. They regard the Shah with awe, naming him as the *khuda-yi kuchik*, or little God, in contrast to Allah who is the *khuda-yi bozorg*, or great God. Village life is very orderly, with most disputes between individuals confined to heated discussions while more serious matters are settled by the headman of the village. There is little robbery within the villages and none of the vices of urban life. Probably the only destructive feature has been the practice of opium smoking. Opium was a means of escape from the monotony of daily toil, but it was more important as a method of alleviating the pain of diseases and sickness for which no medical treatment was available.

In village life there is no distinction between the hours of toil and those of amusement and recreation. There is no organization of leisure hour activity as in the West, and indeed there is little leisure except in the winter months. Conversation is the chief form of diversion and relaxation, when groups of men gather to drink tea, to discuss the weather, and to exchange news and opinions. These meetings also help to perpetuate the stories, song, and ballads of the region and to keep alive the love of poetry which is so characteristic of the Persian people.

The villagers have only limited contact with the outside world, when the men travel to the larger towns with donkey loads of straw, fuel, fruit, and vegetables or when at less frequent intervals the families go to town to shop in the bazaars, or make pilgrimages to the tombs of local saints.

This picture of village life remains accurate for thousands of the smaller settlements, but has been much altered in as many others, as well as in all of the larger villages. The vast program of land reform, accompanied by new facilities and services, has ushered in a fresh era of rural life.

For many long years only about one-sixth of the peasants, *ra'iyat*, or *dehgan*, owned their own land. Such a cultivator was called *khurdeh malik*, or owner of a small piece. There were three major categories of land on which the other peasants worked: *vaqf*, or endowed land; *khaliseh*, or land owned by the state; and *arbabi*, or *amlak*, lands under private ownership. In addition, Reza Shah acquired vast tracts of land throughout the country, which were registered in his name. Much of the *vaqf* land was held by religious shrines: the shrine of the Imam Reza at Mashhad owned a goodly part of the province of Khorasan. *Khaliseh* lands were operated or leased by agencies of the state.

Most of the irrigated land fell into the *arbabi* category, and on much of these areas the peasants were held in comparative serfdom. *Rab*, the word for lord, master, or possessor, has as its plural *arbab*, which may be translated as lord of lords: its parallel to the term shah of shahs is not without significance.

Statistics on the ownership of arbabi land were incomplete and confusing. Some such holdings were known to be enormous, and it was said that twenty-seven families owned 20,000 villages. One landlord, Mehdi Batmangelich, boasted that his holdings were as large as all Switzerland. Making a random selection of some important landowning families to indicate areas of the country where large holdings occurred, one can cite the Ardalan family of Sanandaj, the Afshar in Azerbaijan, the Alam in Birjand, the Amini family in Gilan, the Bayat and Beyklik families in Arak, the Khalatbari in Mazanderan, the Qavam in Fars, and the Qaraqozlu at Hamadan.

These landlords were not a stable landed aristocracy which transmitted its holdings unbroken from generation to generation. On the one hand, the Islamic law of inheritance made the division of property among heirs almost obligatory, and, on the other, years of insecurity and warfare operated to separate proprietors from their land. A few vast holdings had been in the same families for nearly two hundred years, but most seemed to have been acquired in the nineteenth century. Some were acquired by the newly-prosperous groups of merchants and contractors. More recent acquisitions were less for the economic value of land in the form of annual revenues than for the social prestige attached to ownership, and to the possibility of membership in the Majlis. Thus, it resulted that in many regions landowning families were rivals for seats in the Majlis, since a deputy was able to protect and foster the interests of his extended family. Most of the large proprietors were absentee landlords, whose estates were managed by resident *mubashirs*, or agents, and who came infrequently from Tehran to look over their property.

Peasants working for landlords lived in houses of the villages which were owned by these landlords along with the surrounding fields, and were sharecroppers under a system which may be many centuries old. This system was based upon the five items essential to the growth and harvest of the crops: land, water for irrigation, seeds, draft animals and primitive equipment, and human labor. Normally the landlord furnished the first two items and received two-fifths of the harvested crop. In poorer areas he might supply everything except the labor, in which case the share of the farmer was barely sufficient to feed his family. However, the systems of dividing the crops did vary from region to region. In the Khamseh area the farmer with his own oxen and seed received four-fifths of the harvest, and in the Isfahan region the man who contributed only his labor was entitled to one-third of the crop. The produce of the orchards was also divided, but trees planted by the sharecropper became his personal property.

Land taxes have varied considerably in modern times. In 1922 they were 10 percent of the landlord's net share of the harvest, in 1926 3 percent of the gross product before division, and in 1930 8 percent of the gross product. In 1934 a law provided for a tax of 3 percent in kind on all farm produce which entered the markets; this law served to transfer much of the tax burden to the peasant. As recently as 1948 new taxes were imposed on the value of the land itself.

In September 1941 Muhammad Reza Shah ceded to the state all the lands which he had inherited from his father, and between 1942 and 1947 many of them were returned to their original owners. In 1949 a bill provided that all such lands that had not been successfully reclaimed and all royal lands which had not been contested were to revert to the Shah, and to constitute a vaqf whose income would be used by the Imperial Organization for Social Services.

In January 1951 the ruler announced that he would distribute all these lands among the sharecroppers who were working them: these holdings comprised more than 520 villages. He then provided the capital for the Bank of Development and Rural Cooperatives (*Bank Omran*), which undertook the distribution of the lands. They were appraised, the value reduced by 20 percent, and then sold to the new owners on the basis of twenty-five annual installments. The bank used its capital and the funds from payments to establish cooperatives, to construct houses, and provide training in modern agricultural practices. Dr. Mossadeq halted the distribution as one of his actions taken against the position and prestige of the ruler, and instituted action to transfer title of these lands to the state. After his removal the program continued and in the spring of 1962 the last of the royal lands were in the hands of their new owners.

The Farmer's Share Law of 1953 provided that the landlord's share of the harvest was to be cut by 20 percent: 10 percent was a deduction of the portion they received from the sharecroppers, and 10 percent was to be a contribution toward the financing of village councils, which would undertake local self-help programs. In 1956 this law was replaced by the Rural Community Bill which fixed the amount to be collected from the landlords for the village councils at 5 percent of their net income. The responsibility for establishing and managing these councils was given to the Community Development Institute of the Ministry of the Interior. Funds provided by the USOM (the Point Four Mission to Iran) to the Near East Foundation—an American organization which had been active in agricultural demonstration programs in Iran for more than a decade—enabled it to carry out activities on behalf of the Community Development Institute, and by 1957 some 17,000 village councils had been established. These bodies undertook to build schools and baths, improve irrigation systems, and construct roads. An auxiliary service, the Community Development Block Program, trained village-level workers, *dehyars*, to act as local leaders.

Measures were also taken to sell khaliseh lands to peasants. The Agricultural Bank was charged with this program, selling the lands for twenty-five annual payments, and using its funds to set up cooperatives and finance the cultivators.

In 1958 Muhammad Reza Pahlavi began to speak out against the large landowners who had refused to follow his example, even when he had spoken directly to them, and sell portions of their holdings to their tenants. He recalled that he had raised this subject as early as 1942: he declared that he had decided to limit the size of individual landholdings, with the former owners reimbursed by sums which they could invest profitably in productive enterprises. As a result, in December 1959 the government submitted to the Parliament a proposed law, "Law on the Limitation and Reform of Landed Property," which contained some thirty-five lengthy and complicated articles. There was vocal opposition from the religious leaders who declared that the proposed law was contrary to the principles of Islam, and sullen opposition from the landlords. In May 1960 both houses of the Parliament passed a somewhat watered-down version of the original draft. Implementation of its provisions were delayed because of continuing opposition, and near the end of 1961 the Shah instructed the government of Dr. 'Ali Amini to move ahead, and an amended draft was prepared which contained provisions less favorable to the landlords than those of the law passed by Parliament.

This new bill came into effect on January 9, 1962, without legislative approval, since the Parliament was not in session. According to its terms, a landlord could retain any single village of his holdings, or retain title to as many as six *dangs* in separate villages. (The ownership of a village is normally based upon six *dangs*, with the Persian word having the meaning of a sixth part of a piece of real estate.) Within forty days after the effective date of the law, all landowners with holdings in excess of the limit were required to select the property they chose to retain. Children under the care of a head of a family who was also a landlord were entitled to retain the equivalent of one village. Lands excluded from the terms of this law included orchards, tea plantations, forests, and all lands worked by mechanical equipment, i.e., agricultural machinery. All the excess holdings were to be purchased by the state through its Land Reform Organization at figures equal to ten times their current annual revenues, with payments made to the former owners in ten annual installments. These excess holdings were then to be sold to the peasants of those same villages, or those living in the vicinity, at prices ten percent above those at which they were purchased from the former owners with the differential being used to develop irrigation facilities and other agricultural improvements. The new owners were to pay for their land in fifteen annual installments to the Agricultural Bank.

Division of the lands began within a very few months, headed by the very energetic Minister of Agriculture, Dr. Hasan Arsanjani. He stated that the abolition of the landlord-tenant system had long been the ardent desire of all liberals and reformists in Iran, and now that "feudalism is being abolished, the middle class grows, and a democracy of the bourgeoise will be established." He estimated the cost of acquiring all the lands eligible for distribution (about 25,000 villages) at the equivalent of $950 million, with funds to come from provisions in the national budgets and from the Plan Organization. It was also made clear that the new proprietors must join cooperatives and thus assume responsibility for managing their own affairs, rather than relying upon a new bureaucracy. However, a sizeable bureaucracy soon grew up under the Land Reform Organization. So-called Additional Articles were added at later dates to clarify, amend, or extend the provisions of the law of January 9, 1962. The shift in the power structure of the country was reflected in the composition of the XXIst Majlis which included a large number of farmers, land reform officials, and civil servants, replacing the large landlords.

The so-called first and second phases of the land distribution program came to an end in September 1971. By that date 16,351 villages had been purchased at a cost of some $132 million, with about 760,000 families receiving land. These lands included properties which did not come under the law of January 9, 1962, but under the Additional Articles: included were lands leased for 30 years by farmers from their owners, vaqf lands leased for 99 years, and lands voluntarily sold by their owners. Finally, on December 15, 1968, the "Law for the Distribution and Sale of Leased Property to

Lessee Farmers," opened the third phase of the program. During this phase an additional 1,311,000 families were to receive their own land, so that by October 1971 there would be no farmer in Iran who did not own his own land. Outside the scope of this program had been the sale of khaliseh lands in the amount of some $20 million.

Fears had been expressed, more privately than publicly, that with many thousands of farmers thrown on their own, aided only by the new cooperatives rather than the long-time paternal supervision of the landlords and their agents, agricultural production would decline. The available data neither confirm nor refute these fears, because of the fact that recent years of drought throughout the country have adversely affected agricultural production.

After early 1962 the establishment of rural cooperative societies went on at a very rapid rate, and by the end of 1972 it was reported that 10,920 cooperatives were serving 48,842 villages. The more recent trend, sponsored by the Ministry of Land Reform and Rural Cooperatives, has been for these cooperative societies to join cooperative unions which operate on a large scale over a larger area. By late 1972 7,971 cooperative societies had merged into 124 cooperative unions. Both societies and unions manage private and communal purchases of feedstock, fertilizers, and equipment, and the collection, storage, and marketing of agricultural products. There were also in 1973 30 agricultural joint stock companies. In such a company the farmer members turn over their land to be operated by the company, and receive shares in the company in proportion to the amount of land they own. They are paid for their labor by the company and receive their proportional share of its profits.

It may be suggested that the formation of cooperative unions and of agricultural joint stock companies represent definite steps toward the establishment of collective farms. This suggestion seems to be supported by a statement made by the Minister of Agriculture in January 1973, which may be summarized as follows: "If the farmers are to receive the amenities of urban life to which they are entitled and desire, the present pattern of scattered villages with only a limited amount of land and water for each village, makes this goal unattainable. It is planned to set up some 500 large and 1,500 smaller areas where vast areas of land well supplied with water and suitable for agriculture will replace Iran's present 60,000 scattered villages." To comment on this plan, one may ask whether the farmers who have at long last received land of their own will be willing to leave it, and will the enormous efforts to improve these more than 60,000 villages, soon to be described, simply be wasted?

The Literacy Corps, established in 1963, had, by 1972, sent a total of 85,000 young people to work in the villages, primarily as teachers. In 1972 over 17,000 men and women were active in the villages. In 1962 7,000 villages had primary schools, while in 1972 these corpsmen were teaching in 21,300 schools, most of which they had helped to build. Other activities in which they directed voluntary local labor included building and repairing mosques, drilling wells, building or repairing bridges and access roads, building public baths, planting model farms, and planting millions of trees. The female corpsmen also set up libraries and gave classes in cooking, sewing, and housekeeping.

The Health Corps was set up in 1964. By 1972 it comprised 400 medical groups, each one caring for from 20 to 40 villages, bringing curative and preventive health services to the villagers through visits and in the more than 300 clinics built by the corps.

The Extension and Development Corps began its work in 1965. It is made up of men who have completed their studies in engineering, agriculture, and mechanical and electrical work, and their activity in the villages is concerned with modern methods of farming, use of chemical fertilizers, pest control, animal husbandry, and hygienic facilities. Many of these corpsmen had completed post-graduate studies, a remarkable percentage with doctorates in their fields.

The Religious Corps was established in 1971 and began functioning in the following year, with its members sent into the villages as preachers and exponents of the principles of Islam, particularly as they relate to the modern world and Iranian society.

More than 3,000 Houses of Justice have been set up in the villages by the Ministry of Justice. These are local tribunals which handle financial litigation, cases of the disturbance of the peace and family disputes rapidly and without any cost to those involved. By 1972 some 640 rural cultural centers were in operation. In 1973 the first village associations began to be chosen by the secret ballots of the villagers, with members of these associations choosing the village kadkhuda.

Periodicals specifically designed for the villagers include two official magazines, *Rusta* (Villager), and *Ayandegan-i-Rusta* (Future of the Villager), and two private magazines, *Dehqan-i-Azad* (Free Villager), and *Dehgan-i-Ruz* (The Villager of Today).

The annual budget of the Ministry of Cooperatives and Rural Affairs is some $60 million. With considerable additional expenditures made by other ministries and through the Plan Organization it is apparent that the standards of village life display a remarkable improvement.

Towns and Urban Life

Many of the towns of Iran fit into a pattern, populated by more than 50,000 persons and situated about eighty miles apart along the main highways. Most of their sites were chosen because they were originally points of intersection of important trade routes, and some of them serve as collection and distribution points for farming regions. Mashhad and Qum owe much of their importance to the fact that they contain holy shrines visited by a great many pilgrims. Quite a number of the towns were developed as the capital of the country during the Islamic centuries when the capital was shifted from one to another of these: Isfahan, Maragha, Tabriz, Sultaniya, Qazvin, Mashhad, Shiraz, and Tehran. Such towns as Kerman, Yazd, and Damghan were the seats of important local rulers. Many of the smaller ones were "main street" towns which grew up along both sides of a major route to provide services to travellers.

In earlier times, with this arrangement surviving until recent years, the core of the larger town comprised a citadel which housed the ruler, or his local representative, and a military garrison, along with the Masjid-i-Jami', the principal mosque. Frequently the citadel and the mosque were connected by a covered bazaar. The rest of the town was a maze of narrow streets and lanes, and on the outskirts of the towns were the caravanserais. The inhabitants of these towns included officials, merchants, landowners, the servants of these classes, and, above all, craftsmen. These craftsmen produced the elegant items demanded by the royal courts and the local lords, and produced those necessary for the daily life of the rest of the population. Towns became noted for their specialties, and it is of interest to note that these items are still produced with high levels of technical skill. Isfahan, Qum, Kashan, and Kerman weave fine rugs; Isfahan produces *kalamkars*, colorful block-printed calicoes; Qum turns out colored glass; Tabriz is noted for its silverwork; Mashhad for turquoises; Abadeh for spoons and boxes of pear wood; and, Shiraz for *khatem kari*, boxes and furniture with decorative inlay patterns. All these towns bears testimony to the program of modernization begun in the reign of Reza Shah. Beginning about 1930, wide avenues were cut through crowded residential quarters, and helped relieve congestion in the towns and to provide space for the new buildings. In only a few cases have the new avenues destroyed some of a town's former charm; at Shiraz the fine vaulted bazaar was cut through and one of the old palaces was destroyed.

In this stage of the modernization of the urban centers they were provided with new primary and secondary schools, branches of the National Bank of Iran, hospitals, cinemas, and office structures for the various ministries of the state. On their outskirts were located grain silos, electric power plants, oil storage tanks, and factories. The newly-built avenues were increasingly lined with shops which tended to compete for customers with the bazaars. Muddy streets were replaced by asphalted avenues, and piped water was a feature of all towns of any size. This initial stage of development has been surpassed by more impressive progress, resulting in part from the planned decentralization of industry and services. Modern hotels feature the larger towns, and the state has erected inns and motels throughout the country.

In spite of the fact that shops line the avenues a good deal of the retail and nearly all of the wholesale business takes place in the bazaar. Essentially the bazaar is a single long street, from fifteen to twenty-five feet wide, which is covered for its entire length by vaults of fired brick so that merchants and shoppers have complete protection from the summer sun and winter rains. It is lined with small shops, usually about fifteen feet wide and twenty feet deep. At intervals wide portals lead into structures resembling caravanserais, each with a large central court surrounded on all sides by offices or storerooms and, in the finer bazaars, roofed with vaults of fired brick. These buildings are the headquarters of the wholesale merchants, and the courts are usually crammed with piles of rugs, skins, and hides; bales of cotton and cloth; boxes of spices, and other merchandise. The bazaar entrance is on the main square of the town, and the bazaar itself may run for as much as a mile until it dies out in a residential quarter. Often it pursues a winding course, and the larger bazaars have secondary branches parallel with and at right angles to the main street. The shops are stocked with both local and imported merchandise, and sales are made by bargaining rather than by fixed prices. However, both the merchant and the shopper are familiar with the current price of any article, and the give and take of the bargaining process is enjoyed by both parties.

The public baths of the towns are considerably more elegant than those of the villages. Some are recently constructed, but many are more than a hundred years old. The older baths are as much as thirty feet below the street level so that clear, fresh water from underground channels or wells can flow directly into their tanks. The floors are paved with glazed tiles, the lower walls lined with marble or alabaster, and the upper walls and vaulted ceilings are plastered and decorated with painted designs. Among the series of rooms for the various stages of the cleansing process is usually one spacious hall which serves as a general lounging and gossip center. A few sentences from an account of a bath written in the eighteenth century give a still valid description: "Immediately within the Porch is the greatest Cell, or rather a large Room, where they doff their Cloaths, and being undressed leave their Garments; in the midst of this Place is a cistern of cold Water coming into it by several pipes. All the other Cells are so conveniently planned, that every one may breathe a different Air as to degrees of Heat; for some want a Hot Bath, others Tepid, and others a cold Bath. The Pavements are all Marble, on which, the more Hot Water is thrown, the more it increases the Heat, although at the same time the Subterranean Fire be as Hot as it can be. On these Marble Floors they at last extend themselves, when they think they have tarried in long enough, that the Barbers, whose business it is, should wind and turn every Limb and Joint of the Body, before, behind, and on every side, with that Dexterity and Slight, that it is admirable to behold them perform it; whereby they leave no Muscle, Nerve, or Superficial Joint either unmoved or not rubbed."

In the more crowded residential quarters the houses are very similar in plan and construction to those of the villages. Those of wealthy people may be considerably larger and built of fired brick, but the open court and the reception room, opening to the south and sheltered by a porch, remain the principal elements of the plan. The best of the houses are usually on the outskirts of the town in the midst of spacious gardens. To the north of Tehran thousands of apartment buildings have sprung up in the last decade and this same trend is reflected in the other large towns. Amazing to the visitor is the rapidity with which the gardens of these new buildings, established on barren ground, become luxuriant within three or four years.

As noted earlier, the townspeople wear western style clothing. Before the reign of Reza Shah the women of the towns wore over their usual clothing a length of black cloth, called the *chador*, which covered the entire figure and was draped over the head and held in one hand to cover all the face except the eyes. After 1936 the veil was discarded, and the head left uncovered or crowned by a western style hat. After 1941 there was a relaxation against the prohibition of the chador, and women from the poorer families took up the fashion of wearing a length of figured cotton cloth which was carried over the head. So clad the women go out to shop without hav-

20. Aerial view of the Iranian petroleum installations at Abadan

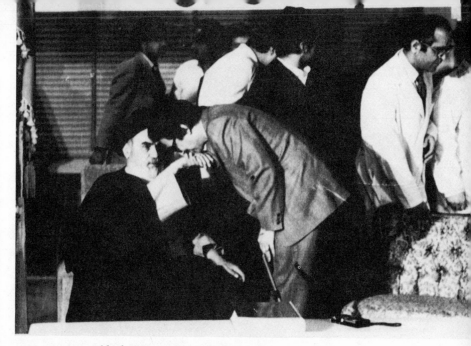

21. Abul Hasan Bani Sadr kissing
the hand of the Ayatullah Khomeini

22. The funeral of the Ayatullah Beheshti

ing to wear their better dresses. Stylishly dressed women do abound on the streets. There are many dress shops to suit all tastes and a number of women's magazines devoted to personal adornment.

Persian women enjoy a greater degree of freedom and equality than their sisters in some Muslim countries. Reference has already been made to the family protection law and to the very marked decline in polygamy. They enjoy the franchise and property rights, and are increasingly active in the professions, and in practically every other occupation. In 1966 numerous women's associations came together in the Organization of Iranian Women which continues to work for full equality in every sphere of private and public life.

Children are the objects of most tender parental affection. Boys are especially favored, but not so heavily as in the earlier days when the birth of a girl was an occasion for condolences rather than for congratulations. This attitude was caused by the inferior status of women; socially, even though the Muslim religion had given her a higher standing than before, and economically, in that the boy was more essential to the family and would eventually take over its support. Mothers still protect their babies and young children against the evil eye, and against sickness and disease, with superstitious rites and charms. The boys become independent and self-reliant at a very early age, and are quite capable of running a shop in the bazaar or working at a trade.

Earlier in this century the cost of living in Iran was very low but the impact of World War II on the local economy set off an inflationary spiral which is still rising. The original cost of living index which began with the base value of 100 for 1936 was adjusted in 1959, 1963, and 1969, each time to a new base value of 100. The continuing inflationary trend is reflected in the following table:

1936	6.6	1960	108.0
1940	10.7	1963	100.0
1945	58.1	1965	115.5
1950	51.6	1969	100.0
1955	77.1	1970	127.0
1959	100.0	1972	112.4

It will be noted, however, that this trend has been slowed down. The cost of housing has risen more rapidly than other items represented in the general index due to speculation in land and buildings in the larger towns. Food costs have been relatively stable, until most recent years when years of drought have lowered agricultural production.

In 1972 the expenditures of the average urban family was as follows:

		percent
	Foodstuffs	43.24
	Clothing	12.50
	Housing	12.06
	Furniture	6.37
	Household goods and services	6.86
	Transportation	5.08
	Health and sanitation	4.84
	Personal care	3.29
	Other goods and services	5.76

The diet of the townspeople is considerably more elaborate than that of the villagers. Rice is the main course of many meals and as cooked in Iran is never moist and sticky, but fluffy and dry and altogether delicious.

When rice is prepared with meat, vegetables, or spices it is called *pelow*. Other foods may be mixed with the cooked rice, or sauces poured over it. One such rice sauce, usually served with boiled chicken, contains almonds, pistachios, orange peel, and dates. Another rice dish contains pieces of fried smoked fish and a mixture of chopped parsley, leeks, dill, and coriander, and a delicious sweet-sour sauce is made of large pieces of duck, pomegranate juice and ground walnut meats.

Kababs or pieces of lamb roasted on a spit are also a favorite article of diet. A popular thick soup, called *awsh*, contains spinach, beet greens, peas, beans, or lentils, and often contains *mast*. Fine omelets are made with greens and herbs.

There is less emphasis on desserts than in our country but during the proper season a great deal of fruit is eaten. The Iranian melons, of which there are at least twenty distinctive varieties including the familiar watermelons, have been famous for centuries. As soon as the season arrives everyone begins eating melons and, according to one old account: "They will eat at that time a matter of ten or twelve Pound of Melon a day, for a Fortnight or three weeks together; and this is as much for Health's sake as it is to please the Palates, for they look upon it as a great refresher and colorer of their blood. After the first Melons, there come up different sorts every Day, and the later these Fruits are, the better. The latest of them all are White, and you would swear that they were nothing but one entire Lump of pure Sugar." Isfahan in particular is noted for its melons, and in one small area north of the city grows a much relished variety which is so fragile and delicate that if a horseman gallops by the fruits will split open on the vine. During the seventeenth century, melons were sent all the way from Iran to the court at Agra in India. They were carried by men who suspended two baskets, with a melon in each, hung on a pole over their shoulders, and walked for eighty days to reach their destination. Nowadays, in the late fall melons are placed in a bed of straw within small caves where they keep in excellent condition until the spring.

Sherbets are still very popular, although apparently less so than they used to be. The word itself is a Persian one (*sharbat*) which has passed directly into the English language. Sherbets were either drinks cooled with ice or snow or the same ingredients frozen into flavored ices: "Sherbets are made of almost all Tart pleasing Fruits as the Juice of the Pomegranates, Lemons, Citrons (limes), and Oranges which are brought to the Markets." One popular drink was made of violets, vinegar, and pomegranate juice and another consisted of water, sugar, lime juice, and a touch of garlic juice. A currently common warm weather drink is made of water, vinegar, sugar, and sliced cucumbers. In the earlier days such drinks were served in huge porcelain bowls and sipped from the large spoons with beautifully carved handles which were the specialty of the wood-carvers in the town of Abadeh, where such carving is still done. However, carbonated soft drinks are now enormously popular: foreign-owned companies bottle cola drinks, ginger ale, and other beverages at

Tehran. Also, there is a very considerable local production of good red and white wines, vodka, and beer.

The large towns have mechanical ice plants but throughout most of the country there are only the local ice houses, whose most important feature is a mud wall at least ten yards high and running in an east-west direction. On the north side of the wall several square pits are dug, and when freezing weather arrives water is put into the pits. Usually the cold will be sufficient to freeze only an inch or two of water during the night, and each day the accumulated ice is broken up and more water added while the high wall prevents the sun's rays from melting the ice during the daytime. After a week or so when each piece of ice is five or six feet thick, it is removed to a special storage chamber under a roof thickly insulated with brush or reeds. Snow is also commonly used during the summer, collected and stored by dwellers in the highest mountain villages, and when hot weather comes, carried on donkey back to the nearest town.

The townspeople normally have a better education and higher social position than the villagers, and are the products of long centuries of cultural continuity. Social graces have been developed to a very high degree. As far back as the seventeenth century a visitor to the country wrote: "The Persians are the most Civilized People of the East, and the greatest Complimenters in the World." Another traveler remarked: "Friendly and Courteous Salutation is no where so much promoted as among the Persians, for the very Plebians, in other parts surly and unconversable, are here Affable and Kind, not Rude and Unmannerly."

The many forms of polite address have grown into a formal ritual of almost endless variety. Such ingrained habits of speech are not easily cast aside, and today the people, including even those who endorse a complete break with the past, begin speaking with such phrases as "I make the humble remark," or, "This slave believes that . . . ," while the phrase "you and I" will be rendered in Persian as, "This slave and your honor." The phraseology is, however, less ornate than in former years. To say a man was dead the speaker used to say, "He has made a gift of the share of life which he had, otherwise he might have lived still many years; but out of the love he has for you he has joined them to those you still have to run."

A characteristic form of Persian politeness was the use of honorary titles. When these were conferred by the Shah on his generals and nobles, they often took the place of personal names. Common titles meant: "Upholder of the Realm," "Intelligence of the Empire," and "Splendor of the Country." Under Reza Shah a law was passed which forbade their granting and use, but cabinet ministers and members of the older generation are sometimes still known by their titles.

Certain forms of entertainment more common in former days still linger on both in the older residential quarters of Tehran and in the towns. The magician makes his rounds, and the fortuneteller attracts his clients. The snake-charmers put on a show which includes, strangely enough, an educational lecture on the life and behavior of various species of snake. Dervishes in their picturesque costumes were much in evidence until about fifteen years ago, when the authorities decided that they were out of place in the modern picture and forbade them to enter the towns. Now they have begun to appear once more. The word dervish is itself a Persian one, meaning "humble," "poor," or "one who lives by alms." The dervishes really correspond to our Christian monks, and are divided into a number of religious orders each with its own tenets and regulations. In Iran the dervishes lead a more wandering life than in the other Muslim countries. A minority are deeply religious men who have taken vows of poverty and chastity, but the majority have lived by their wits and on the credulity of the common people. Their usual costume is a long varicolored patchwork cloak belted with a rope girdle from which hangs a string of wooden beads. Over their shoulders is draped a dressed panther or wolfskin, and on their heads is a skullcap embroidered with verses from the Qoran. Their matted hair hangs to shoulder length and their feet are often bare. Slung over one shoulder is a horn and they carry a weapon for protection against wild beasts, usually a steel battle-axe or a wooden club studded with sharp spikes, and a begging bowl made from a hollowed gourd and decorated with fine carving.

The *zur khaneh* or "house of strength" is an old Iranian institution which continues to flourish both in Tehran and in the provinces. Its organization is like that of a physical culture society, and its members are drawn from every occupation and social level. It is believed that the zur khaneh originated in a remote period when Iran was occupied by foreigners and the youth of the country trained in secret against the day when they would be able to expel the invaders. Each zur khaneh has one large room containing a pit about twenty feet square and three feet deep, a raised platform for the drummer, and a space for spectators. About twelve men, clad only in gaily embroidered knee-length trousers, enter the pit and perform to an accompaniment of stirring drum beats and the chanting of verses from the *Shah nama*; they do push-ups in unison, juggle large and very heavy Indian clubs, and jump high in the air and spin about. The military origin of the zur khaneh seems to be attested to by exercises with large wooden shields and iron bows which have links of iron chain in place of a bow string. Wrestling matches, in both Persian and European style, are the climax and finale of each session.

Tea houses and cafés are the popular public gathering places of the small towns where men drop in for tea, for friendly conversation, for a game of backgammon, and to smoke the water-pipe. Tea is the Persian's favorite drink: it is served in glasses with plenty of sugar, although often the drinker holds a lump of sugar between his teeth and sips the tea through it. The custom of tea drinking seems to have come from Russia fairly recently, for during the seventeenth and eighteenth centuries the Persians drank a great deal of coffee, which was quite unknown in the West until European travelers discovered it in Turkey and Iran. In 1652 a Mr. D. Edwards, who was a merchant trading with Turkey, began to call himself a "coffeeman" and to import the berries to England, where coffee drinking soon became a fashionable fad. An Englishman who visited the coffee houses of Isfahan in 1627 wrote: "The coffee, or coho, is a drink black as soot, or rather a broth, seeing that they sip it as hot as their mouth can well suffer out of small China cups. 'Tis strongly scented and somewhat bitter, distrained from berries beat into a powder and boiled with water: if supped hot comforts the brain, helps raw stomachs, aids digestion, expels melancholy and sleep. However unsavory it seems at first, it becomes pleasing and delicious enough by custom."

At Tehran cafés replace the tea houses of the towns. Many of them are in pleasant gardens and in the late afternoon they are crowded with family groups who linger for coffee, tea, a cold drink, ice cream, and pastry. Restaurants are, of course, to be found in every town, including many which specialize in *chelo kabab*, kababs on a bed of rice, along with soft drinks and ice cream. Formerly Persian families almost always ate at home, but this is no longer the case.

One of the most common family diversions is a holiday trip into the country. The weekly holiday of the Muslims is Friday, and on this day the families pack a picnic lunch and set out by car or bus for some shady garden beyond the edge of town.

There are fifteen official public holidays in the year: ten are religious holidays, three mark the advent of the New Year, one celebrates the birthday of Muhammad Reza Shah and the other the anniversary of the constitution.

No Ruz, the New Year, is March 21, the first day of spring and the sun's vernal equinox. The Iranians have celebrated the arrival of spring for many centuries as evidenced by the long rows of sculptured reliefs at Persepolis showing groups of people from every part of the mighty Achaemenid empire bringing their tribute to the ruler of Iran on this occasion.

Preparations for No Ruz begin well in advance. Fifteen days before the festival, each household plants in a shallow bowl seeds of wheat or lentils which by the proper time have sent up a thicket of fresh green shoots several inches high as a token of spring.

For No Ruz household servants and minor government employees receive an extra month's wages, the house receives its spring cleaning, and everyone dresses in new clothes. On New Year's Eve a light must burn in every room in the house and a special table is prepared. The centerpiece consists of a mirror and candlesticks, and grouped around it are a copy of the Qoran, a large sheet of bread, a bowl of water in which floats a green leaf, a glass of rosewater, nuts, fruit, candy, colored eggs, chicken, and fish. A large plate or tray contains the *haft sin*—seven articles whose names begin with the Persian letter S. *Sepand, sib, sir, serkeh, samanu, sabzi,* and *sumaq* are the English wild rue, apples, garlic, vinegar, a paste of malt grain, greens, and sumac.

As the time of the vernal equinox approaches all the household is grouped around the table to await the exact moment of the New Year, said to be marked by the moving of the leaf on the surface of the water or by the rotation of an egg placed on the mirror. In large towns a cannon is fired to announce the New Year. Not all the customs and ceremonies associated with No Ruz are now performed in the towns, but those described below are observed by every family.

The New Year begins a period of five days of official holidays devoted to social calls. During the first two or three days the eldest members of the family remain at home to receive calls from friends and relatives, and large sums are spent on the entertainment of the guests. Return calls are paid, countless greeting cards are mailed, and the atmosphere of these days is one of contentment and rejoicing.

On the thirteenth day of the New Year, considered an unlucky day, the bowls of green shoots grown in the houses are thrown out into the streets, if possible into running water, and everyone troops out into the open country for a promenade in the fresh green fields. Each family takes along as elaborate a supply of food as its means permit, and spends the entire day in the open. The people believe that in this way they do not only welcome the spring but carry the bad luck associated with the thirteenth day away from their homes and abandon it in the country where it can do no harm.

Tehran

Writing early in the twelfth century a geographer mentioned Tehran as a village in the district of Ray. For centuries it remained small and obscure, with a limited reputation for its fine fruit and shady groves. The Safavid rulers camped at the site and in A.D. 1785 Karim Khan Zand had a royal palace erected there. In 1788 the Qajar ruler Aqa Muhammad chose Tehran as the capital of the country: at that time it had fewer than 15,000 people. This ruler began the Gulistan palace and his successor Fath 'Ali Shah built the once vast palace and garden called the Qasr-i-Qajar north of the town in the early nineteenth century and also the Masjid-i-Shah, or Imperial Mosque, still standing in the bazaar area.

Nasr al-Din Shah, 1848-1896, who made several trips to Europe, tried to emulate European towns and provided Tehran with a horse-drawn tramway, factories, administrative structures, a hospital, a power station, post-office, etc. The covered bazaar and the impressive Sepahselar mosque were erected near the end of his reign as were royal palaces in the Shemiran area. The population reached 170,000.

Slowly growing Tehran took a fresh lease of life during the fifteen-year reign of Reza Shah. The old city walls were demolished and an entire new town sprang up on the northern outskirts of the older one. Wide, asphalted avenues, marble-clad buildings for the ministries, a monumental officers' club, and a central police headquarters and a national bank building both modeled after the architectural style of Achaemenid times. Since 1946 this growth has been even more rapid; a few years ago Shah Reza avenue was at the northern limits of the city but now the built-up region of housing and shops stretches far to the north. Industrial plants spread over areas to the south and west of the capital, and municipal housing developments have grown up in the suburbs.

Several wide avenues climb some six miles to the Shemiran area, several hundred feet higher than Tehran and much cooler in the summer. Colder in the winter too, and although the city has snowfalls in the winter usually the snow melts fairly soon. Thousands of new homes and apartment buildings, and, more recently, condominiums, are filling up the entire region between the city and Shemiran. The city is well supplied with excellent hotels and restaurants, the latter located in gardens during the summer months. Houses in Shemiran feature spacious gardens and swimming pools.

Most of the shopping takes place to the north of Shah and Istanbul avenues. Visitors are attracted to the colorful rug shops and the many stores, concentrated along Firdawsi avenue, crammed with Persian antiquities and modern handicrafts. Lalezar avenue has its own row of fascinating jewelry shops where pieces may be made to order.

The visitor has a considerable choice of things to see and do. In the city are the Archaeological museum; the Ethnographical museum; Sepahselar mosque; the old bazaar and the Masjid-i-Shah; the extensive Gulistan palace featuring a royal museum with two jewel-encrusted thrones and the imperial collection of illuminated manuscripts and miniatures; and the National Bank with its display of the fabulous crown jewels.

The Shemiran area comprises some 60 villages nestled in the foothills of the Alborz range. The Qajar palace Sahib Qaraniyeh is at Niavaran and just above Tajrish are the extensive grounds and the summer palaces of the present royal family. Throughout the area footpaths along mountain streams wind steadily upward passing falls, pools, orchards and meadows and many charming picnic sites.

Tehran proper has some 3 million people and its immediate environs add another half million. More than do great cities elsewhere, it overshadows all the others in the country in the concentration of administration, industry, services, and business. It contains 28 percent of the total population. Living there are 35 percent of the civil servants, 46 percent of the architects and engineers, and 53 percent of the doctors and dentists of the entire country. It has 70 percent of the educational institutions, 65 percent of the automobiles, and 25 percent of the cinemas. The programs of its television stations are relayed to the other parts of the country.

The capital is beset by the problems of large cities everywhere, and by some special ones. In the latter category, the city lacks a sewage system, and has no subway or elevated line. Transportation is by private cars, taxis, and buses. Traffic jams tend to build up on many of its 12,000 streets: parking lots are few in number, and parking meters a very recent innovation along a few streets. Several hundred buses depart or arrive every day for destinations within the country, and even to Europe. The international airport is so busy that a new one is required, and is in the planning stage.

The budget of the municipality amounts to about $70 million annually, with much of this sum going to maintaining the many spacious parks, and to the construction and upkeep of the streets and avenues. With some 660 factories and 1,800 smaller industrial units, air pollution is an increasing problem. Several master plans for controlling and directing its future growth have been considered, and part of a system of ring roads has been built. The planners believe that by 1990 Tehran will have at least 8 million inhabitants, and at that time the reservoirs which supply water to the city will be inadequate. Current problems include that of prostitution, with some 1,500 prostitutes quartered in the Shahr-i-No, the so-called New City, in southwestern Tehran. Youth gangs engage in juvenile crime, although the streets of the capital are very safe. A small minority of the youth of both sexes dress in blue jeans and army jackets and sport sun glasses and long hair. Grass is smoked, and time passed in cafés and discotheques. Distinctive to Tehran is the practice of young men assuming the names of automobiles assembled in Tehran factories, such as Fiati and Peykani.

With a car the visitor can cover a wider area. Historical remains are to be found at Ray, five miles south of Tehran and once one of the mightiest cities of the mediaeval world. Shah Abdul Azim, just adjacent to the ancient site, has its saintly shrine, crowned by a golden dome, and nearby is the stately mausoleum of Reza Shah. Varamin, twenty-five miles south of Tehran, has monuments of Muslim architecture of the thirteenth and fourteenth centuries—a mosque, tomb tower, and shrine. Demavand village, some 35 miles east and north of Tehran at a height of nearly 7,000 feet, is located in a picturesque mountain setting, dotted with structures of the Seljuq and Mongol periods. In the summer many people pass through Demavand village on their way to climb Mount Demavand; the climb itself is not difficult but sulphur fumes make the experience somewhat trying. In winter skiing is featured at Ab 'Ali, some 30 miles east of Tehran, and elsewhere.

Tehran offers three different faces to the visitor or resident: the crowded area of the old town which has little to attract him; the cosmopolitan, bustling new city, teaming with traffic and with new construction springing up everywhere; and the surrounding countryside with its countless pleasant retreats from the turmoil of the town.

Isfahan

Above all other towns in Iran, Isfahan rewards the tourist in search of the monuments and atmosphere of the country's glorious culture. Less than an hour by plane from Tehran or a short day's run by bus or car, Isfahan is particularly delightful in the spring when flowering orchards are everywhere and the Zayandeh river fills its banks as a noble river should.

The foundation of Isfahan dates from the pre-Islamic centuries and it was a renowned center of trade and culture under local Muslim princes and then the capital of the vast kingdom of the Seljuqs. Spared devastation by the Mongols, it reached its greatest fame under the Safavid rulers and fell into slow decline after siege and capture by an Afghan army early in the eighteenth century.

At the end of the long covered bazaar is the great Masjid-i-Jami', or Congregational Mosque, whose great open court is flanked by dome chambers, ivans, and prayer halls that were erected at intervals over the long centuries. Two dome chambers are of the Seljuq period and feature decoration and inscriptions in brick patterns, while additions of later periods display surfaces of bright faience mosaic.

But the principal spectacle at Isfahan is the remains of the imperial city built by Shah 'Abbas between the old town and the river. Celebrating the New Year festival at Isfahan in 1598 the Safavid ruler decided to move the capital of Iran to this spot. The focal point of the grandiose plan was the Maidan-i-Shah, or Imperial Square, a rectangle 570 yards long and 175 yards wide. A uniform façade in two stories still encloses the square: its shops housed sellers of jewelry, cloth, and drugs. On weekly market days the square was thronged with booths and buyers and on special occasions the Shah and his courtiers played at polo; the stone goal posts stand at either end. Today the shops on the ground floor of the façade around the square display the fascinating handicrafts of Isfahan—pottery, miniature paintings, silverware, and vessels and trays of copper and brass—as well as a larger area where rug weavers may be seen at their looms. The center of the square has been filled with pools, fountains, lawns, flower beds, and electric light poles, and many of those who love Isfahan hope to see all these removed and the great square restored to its original condition.

In 1613 work was begun on the Masjid-i-Shah, or Imperial Mosque, situated at the southern end of the maidan and although the ruler pressed for speed the entire complex was not completed until after twenty years of unremitting effort. The mosque is one of the supreme monuments of Persian architecture. From a lofty entrance portal the way leads into an open court flanked by great ivans and arcades; all the bulk of the mosque is turned at an angle to the square itself in order to be oriented directly toward Mecca. Inside and out, the surfaces of walls, pillars, vaults, minarets and dome are clad in multicolored tiles in which light and dark blues predominate.

A monumental entrance to the old covered bazaar was erected at the northern end of the maidan in 1617. Traces of a huge painting which depicted the victory of Shah 'Abbas over the Uzbeks is still visible on the rear wall. Crowning the portal was a musician's gallery where drums and oboes played long and loud every night just at sundown.

Near the center of the long east side of the maidan is the well preserved mosque of Shaykh Lutfullah. This was one of the first structures to be started, although the decorative details were not completed until 1619. From its brilliantly colored portal a corridor leads around and into the single chamber of the mosque; at a higher level the square chamber is transformed into a lovely dome, pierced by sixteen grilled windows. Every square inch of the interior is decorated with mosaic faience and the light streaming through the grilles pours changing patterns of color upon the sparkling walls. The effect is so elegant, so rich, so dazzling, so unique that the visitor loses touch with time and reality.

On the western side of the maidan stands the Ali Qapu, or Lofty Gateway, the noble entrance to the palace grounds and gardens. In addition to that function it was the administrative center for the royal court, dealing with countless personnel and equipment, supplies and treasures. Rising some six stories, the structure had as many as ten rooms on some levels; a lofty open porch overlooked the entire maidan and eighteen huge tree trunks sheathed with flat strips still uphold the roof decorated in blue and gold. Royal audiences were held at the Ali Qapu and accounts of the arrival of foreign ambassadors with their precious presents make fascinating reading. Inside the structure all the ceilings were decorated with floral patterns in red, blue, and gold, while the walls had scenes of courtly pleasures in country and garden; much of this decoration remains.

Of the many pavilions and palaces scattered through the gardens to the west of the Ali Qapu only the palace called the Chehel Sutun, or Forty Columns, remains. Restored and housing a museum of objects from the Safavid and earlier periods, it is a highlight of a visit to Isfahan. The great open porch with its towering wooden columns overlooks a long reflecting pool and from a recess back of the porch three doors lead into a banquet hall which extends the full width of the structure. Six enormous oil paintings adorn the upper walls and help to visualize the turbulent and lusty life of the seventeenth century: they show battles and royal parties in which dancing and drinking hold the stage.

To the west of the palace area proper was the Chahar Bagh, or Four Gardens, not intended to serve as a busy street as at the present day but as a place of promenade. From a corner of the palace grounds it ran down grade for nearly a mile to the river, crossed by the Allah Verdi Khan bridge, and then up rising ground to a now vanished royal estate called the Hazar Jerib, or Thousand Acres. Eight rows of plane trees and poplars, among which grew a profusion of roses and jasmine, were spaced across the sixty-yard width of the promenade. Five watercourses ran down the avenue and at each change in level was a marble pool with its fountains. Not until near the end of the nineteenth century were many of the fine old trees cut down, while in recent years wide strips at each side have been paved with asphalt. Just enough of the original aspect remains to realize that a seventeenth century traveler was deadly serious when he described it as the most beautiful avenue that he had ever seen or heard talked about.

Both sides of the Chahar Bagh were lined with garden pavilions belonging to nobles and set in spacious grounds: one of these, the Hasht Bihisht, or Eight Paradises, has recently been restored: its domed central room features a pool and fountain, with wide arched openings to the gardens on four sides. Just down the avenue is the Madrasa Mader-i-Shah, or Religious School of the Mother of the Shah, completed in the year 1714. An entrance façade of multicolor tile and polished marble leads into a rectangular court. Two-story arcades once provided a series of rooms for teachers and students and the reflections of these whitewashed arcades shimmer in a dark pool. A great blue tile dome rises high above the sanctuary chamber. Adjacent was a vast caravanserai whose revenues were devoted to the upkeep of the religious school. A few years ago this caravanserai was rebuilt, provided with all modern conveniences, and opened as the Shah 'Abbas Hotel. The interior public rooms are a real museum of the finest Persian handicrafts: tiles, paintings, inlay work, and woodwork. One side of the religious school was flanked by a bazaar covered by a long series of very high vaults. This too has been restored and houses shops which display these same handicrafts, and others.

Crossing the Zayandeh river on the bridge mentioned above one reaches Julfa, the settlement of Armenian weavers and craftsmen brought by Shah 'Abbas from Julfa in Azerbaijan. These settlers enjoyed royal protection and built a splendid cathedral in the years between 1606 and 1654 and two additional churches. Today the cathedral is open and deserves a visit, both as an architectural monument and because of the unique manner in which the Persian tilework was used in the interior. In the grounds of the cathedral is a museum of Armenian antiquities and treasures.

As the visitor tours the monuments of the time of Shah 'Abbas and of other periods he has no way of knowing that years of painstaking effort have gone into the restoration and preservation of these structures. National monuments throughout the country are in charge of the Antiquities Service, and at Isfahan skilled tile cutters, masons, plasterers, and engineers have been at work for a quarter of a century. As one example, in the 1930s the exterior surface of the great domes were in pitiful condition; then they were surrounded by towering scaffolds, and after years of cutting and fitting minute pieces of tile in place these surfaces were restored to their original condition. Wall surfaces received similar treatment. Currently teams of specialists, furnished through UNESCO, are engaged in further exploration, consolidation, and reconstruction of these monuments.

Some suggestions for shopping have been made in earlier paragraphs. Not only is the Chahar Bagh lined with fascinating shops, but no visitor should miss strolling through the almost endless old bazaar where bargaining is expected and enjoyed. Just at its entrance, at the northern end of the maidan, is a government handicraft shop crowded with treasures. In addition, Tehran, Shiraz, and other towns have branches of these shops.

Shiraz

This southern town is renowned for its poets, its gardens, and its wine; all three having centuries of tradition. Just over an hour by air from Tehran, it is a day's journey by car from Isfahan. The trip to the Achaemenid capital of Persepolis only 35 miles to the north is an easy one; excellent hotel accommodations are available at the site.

Shiraz has the long and eventful history common to many Persian towns. For centuries it has been the capital of Fars, or Pars, the area which gave its name to Persia. The town was devastated in turn by Timur, by a great flood, and by the same Afghan army that took Isfahan, but in the middle of the eighteenth century it came back to life and splendor under the benevolent attention of a regent-ruler of Iran, Karim Khan Zand. He embellished the town with a citadel at its center, a magnificent covered bazaar with very high vaults of wide span, the Masjid-i-Vakil, or Mosque of the Regent, his own mausoleum, and spacious gardens on the slopes above the town.

By car one approaches the city from the north and through the pass of Allahu Akbar, "God is Great," so called from the usual expression of wonder at the sight of blooming, fertile Shiraz. The wide avenue leading down into the town was once lined on either side by gardens and pavilions; a surviving structure houses a hotel which offers a splendid view across groves of orange trees and ranks of stately cypresses. Indeed the soaring cypress trees are the hallmark of Shiraz.

Within the town the visitor may see the Masjid-i-Vakil, the earlier Masjid-i-No, or New Mosque, and the Masjid-i-Atiq, or Ancient Mosque. The latter structure has its name from the fact that one section may have been erected as early as the end of the ninth century A.D., but in recent years garish reconstruction and repairs have hidden the work of earlier periods. The soaring, bulbous domes of revered shrines dominate this old section of the town.

A wide avenue was cut through Karim Khan's bazaar, but a very lengthy section remains intact and it remains a fascinating place for shopping, specializing in goods desired by the nomads and in the rugs which they have made. Nearby is an octagonal pavilion set in a garden, believed to have been erected by Karim Khan as his own mausoleum. Few of the townhouse and garden structures have survived in the heart of old Shiraz. Noteworthy is the nineteenth-century Narangestan, which now houses the Asia Institute of Pahlavi University. All the eighteenth- and nineteenth-century buildings in Shiraz feature distinctive tilework in which roses and other flowers predominate.

Outside the town there is much to see. To the northeast are the tombs of Hafiz and Sa'di, places of pilgrimage for all Iranians. Hafiz lies under an open kiosk, surrounded by other graves, and in a fragrant garden. Persian visitors take a *fal*, or augury, at the tomb; his poems opened at random give the answer to questions in the minds of his devotees. Some distance to the east is the tomb of Sa'di. A few years ago the old structure sheltering the grave was replaced by a new building, combining elements of Muslim style and of modern construction.

Around the town are the old gardens, or *baghs*, of Shiraz, each one containing an impressive pavilion, often with many rooms. Most have recently passed from private to public hands. They include the Bagh-i-Eram whose handsomely restored pavilion houses visiting dignitaries, the Bagh-i-Golshan, restored as a residence for the royal family on its visits to Shiraz, and the Bagh-i-Delgosha. Near the Bagh-i-Eram are the spacious grounds and the many buildings of the Pahlavi University, while just across the Rudkhaneh Khushk, the Dry River, is the imposing Nemazee Medical Center.

The inhabitants of Shiraz appear to take a pride in their city, its history, and its modern features not found in the other larger towns. Local philanthropists have been numerous and generous, while serious efforts have been made to preserve its abiding charm.

Tabriz

Tabriz, capital of the province of eastern Azerbaijan, is situated in northwestern Iran. It is 370 miles by paved highway from Tehran and one hour by air, and is also a station of the railway line which runs from Tehran to Istanbul and on to Europe. Long one of the leading commercial centers of the country, located on the great highway across Asia, it was also the capital of the region during the Mongol period and again in Safavid times. Until this century the crown prince had the post of governor-general of Azerbaijan. Rainfall is more abundant in this region than over most of the plateau and dry farming is customary; the area is an abundant granary.

Tabriz itself has few monuments of great interest. The remains of the huge vaulted Mosque of 'Ali Shah, built early in the fourteenth century and later incorporated into the citadel are visible, and the restored Blue Mosque of the fifteenth century displays the finest mosaic faience and carved alabaster.

Mashhad

Rivaling Tabriz in commercial importance but far more noted for its holy shrine, Mashhad is an hour's plane flight from Tehran and at the end of a branch line of the Iranian railway which traverses the northeastern part of the country. Long the capital of the province of Khorasan, Mashhad—the very name means "burial place of a martyr"—owes its chief fame to being the site of the burial shrine of 'Ali Reza, eighth Imam of the Shi'a line. The Imam Reza was poisoned in the ninth century.

The town is bisected by a north-south avenue and at its heart a circular avenue encompasses the shrine on all sides. Great open courts flank that part of the shrine which contains the actual tomb chamber. That to the west is the oldest and is represented by the Mosque of Gawhar Shad, built between 1405 and 1417. That to the east is largely of the Safavid period and various Safavid and Qajar rulers supplied the funds for gilding the great dome, its flanking minarets and the arched portal nearest the tomb chamber. The chamber itself has lower walls encased in tiles and upper walls and ceiling entirely covered with mirror work. The great sarcophagus, set in one corner, is sheltered within silver grillwork. Throughout the years streams of pilgrims come to the shrine from all corners of Iran, and from Pakistan, Afghanistan, and Iraq. Long inaccessible to non-Muslims, the Iran National Tourist Organization insists that it is open to other visitors who have made prior arrangements: in fact, this is not always the case, particularly at the time of Muslim religious holidays.

The finest treasures of the shrine—carpets, hangings, bronzes, manuscripts, miniature paintings and illuminated Qorans—are housed in a modern museum building which is open to all visitors. Among modern products the fine carpets of the province and the turquoises, from mines in the vicinity which have been worked for centuries, are outstanding, and may be purchased in the busy bazaars which edge the shrine itself.

Education

Iran's history of education goes back to a very remote period. In Achaemenid times the young men were taught not only to ride and to shoot the bow, but to know the value of truth and to distinguish between good and evil. After the Arab invasion and the adoption of the Muslim religion, education was based upon the Qoran just as in Europe it was based upon the Bible. For many centuries Muslim priests taught, in return for a very small sum of money from each parent, in schools called *maktab*, where the children memorized the Qoran by chanting its verses in unison and learned to read and write Persian and do simple arithmetic. There were also many religious colleges, something like the western theological seminaries, where advanced students, gathered around men renowned for their learning, worked at such subjects as the interpretation of the Qoran, religious law, and religious philosophy. There were no formal examinations. No place for girls was provided in this system of education.

Fairly close contact with some of the European countries was established in the early seventeenth century, and from this time on interest grew in foreign ideas, languages, and way of life, culminating in the establishment in the nineteenth century of schools modeled on western ones. One of these was the *Dar al-Fonun*, or House of Learning college, founded at Tehran in 1852, which at first specialized in military subjects, but soon broadened into the field of the liberal arts and performed a vital service in educating the young men of the leading families. The first Ministry of Education was founded in 1855, but the essential form of the present educational system dates from the organization of an Education Council in 1897, when the decision was taken to use the French system of education as a model.

However, Western education had reached Iran earlier through the American schools conducted by the Presbyterian Board of Foreign Missions. After the first school opened at Rezaieh (then called Urmiya) in 1836, others, attended largely by Assyrians and Armenians, were organized in the vicinity. Schools of higher grade for both boys and girls were gradually established by the Americans in other cities, and were attended by an increasingly larger proportion of Persians. A school for boys was established in Hamadan in 1870 and one for girls twelve years later. In Tehran an elementary school for boys was opened in 1873 and one for girls the following year. At about the same time schools were opened in Tabriz, and a school at Rasht developed into separate schools for boys and girls in 1907. The last American school was opened in Mashhad in 1926. At Tehran Alborz College, the outgrowth of the American High School for boys, in 1928 received a temporary charter from the Board of Regents of the University of the State of New York and in 1932 this charter was made permanent. The development of this institution was the life work of the late Samuel Jordan, principal of the school from 1899 and president from the time it became a college until his retirement in 1940.

At the same time the Sage College for Women was being developed as a continuation of the Nurbakhsh School for Girls, operated in relationship to Alborz College. In addition to these American schools there were British, German, French, and Russian schools in operation in various parts of the country. In 1940 all foreign-run schools which were educating Persian students were taken over by the Iranian government; a private explanation of this action was that it was taken to halt the spread of Soviet propaganda through the media of the Russian schools. This action was later rescinded.

In the current annual budgets of the government some 14 percent of revenues go to the Ministry of Education and the Ministry of Science and Higher Education. An additional 3.5 percent represents subsidies to the universities of the country.

In 1943 the Majlis passed a law making education compulsory for all children between 6 and 12 years of age. Although the law has never been fully implemented, year by year it is imposed on designated areas. The school cycle has comprised two six-year periods of primary and secondary education. In 1965 a new cycle was introduced, as reflected in the accompanying diagram. This cycle is put into practice area by

Simple Vocational Training

Technical Colleges

Colleges and Universities

Technical Education (Training of Technicians)

Vocational Education (Training of skilled worker)

Academic Secondary Education

Educational Guidance (Second Cycle of General Education)

Primary Education (First Cycle of General Education)

Kindergarten

THE NEW CYCLES OF EDUCATION

area, and its purpose is to provide directed choices between continuing study in either the technical or academic fields. Kindergartens, or *kudakistans*, are relatively few in number. In 1972 there were 18,396 primary schools, or *dabistans*, functioning under the Ministry of Education, of which number 1,256 were private schools. They were attended by 2,888,000 children, of whom more than half lived in urban areas. In addition, the Literacy Corps was operating over 10,000 rural primary schools, with about 320,000 pupils.

Characteristic of changing methods of education are the readers used in the primary schools. Formerly, the second reader contained short articles about being careful in the streets, when filling the oil lamp and when playing near the edge of the courtyard pool; animal fables; lessons on plants, animals, and geography; and simple stories based on older history, myth, and poetry. By non-Persian standards the amount of poetry in this reader was very high. About ten years ago a group of educational specialists was set up to develop new readers, and those now in use much resemble those familiar in the United States: boys and girls discuss their activites on the private and public scene, and the readers are well printed and beautifully illustrated in color. In the government schools primary education is free, as are the readers and other school materials.

In 1972 the secondary schools, or *dabiristans*, included 2,600 government schools and an unstated number of private schools, with a total enrollment of 1,134,000 students. These secondary schools have demonstrated a remarkable growth, with almost three times as many students in 1972 as there were ten years earlier. In the usual six-year cycle there is a general examination at the end of the third year, and in the later years the students may concentrate upon literature, mathematics, commercial subjects, or the natural sciences. As many as seventeen subjects were studied from one to four hours a week, but the trend has been to reduce this number. Of the foreign languages, Arabic is stressed since it is the language of the Qoran, while English has far outstripped French in the role of the most popular foreign language.

The University of Tehran was founded in 1935 on the

orders of Reza Shah and was operated by the Ministry of Education until 1943 when it was granted autonomous status. The largest of the country's universities, it has seventeen schools, or colleges. Statistics relating to the universities, or *daneshgahs*, are given in the following table.

THE UNIVERSITIES OF IRAN IN THE ACADEMIC YEAR 1970-71

Name	Founded	Students	Faculty	Medical School
Tehran	1935	17,305	764	Yes
Tabriz	1947	5,187	315	Yes
Isfahan	1950	3,654	154	Yes
Jondi Shapur, Ahwaz	1955	2,099	101	Yes
Firdawsi, Mashhad	1956	3,075	164	Yes
National, Tehran	1960	6,106	220	Yes
Pahlavi, Shiraz	1962	3,483	311	Yes
Aryamehr Industrial, Tehran and Isfahan	1966	1,746	192	No
		42,655	2,221	

Pahlavi University was established by order of Muhammad Reza Shah to counterbalance the French system of instruction which still prevails at the University of Tehran, with its emphasis on the accumulation and memorization of facts. Instruction is primarily in English: it has a number of American teachers and ties with American universities. Jondi Shapur University is in the region where there was a famous intellectual center and medical school active in the Sasanian period and into the Islamic period. The National University was the only private university functioning in 1970-71: its very modern campus is in the Shemiran area. In 1970-71 the Aryamehr Industrial University was in the process of moving from Tehran to Isfahan, because of the location of the steel mill and other industrial enterprises near Isfahan.

Under three-fourths of the university students major in the social sciences and the humanities, with lesser numbers in engineering and medicine, and still smaller numbers in other fields. Country-wide university entrance examinations are held in July. In 1972 some 80,000 applicants took this examination, and over 10,000 were accepted for admission in the universities that autumn. Obviously, the existing universities need to take in more applicants and new universities are required. Kerman University was expected to be established in 1973, and Hamadan University in 1974. Also, 1974 was the anticipated date of the opening of the Iran-France University, with an initial enrollment of 3,000 students. With instruction primarily in French, the faculty will be in part French and in part Persians educated in France. In addition, the Ministry of Science and Higher Education plans to establish a so-called T.V. University and a Labor University.

Remarkable headway has been made in setting up the vocational and technical schools which supply the skilled manpower to staff Iran's burgeoning industry. Such schools have long been in existence, but not on the present scale. In 1972 there were 52,000 students in well over 300 schools. Tehran has a Polytechnic School and an Institute of Technology. One of the newest is the Nafisy Teknikom at Tehran which provides training in the fields of electric power, mechanics, road building, and construction. Iran and the USSR have an agreement which provides for Soviet cooperation in establishing industrial training institutes. Such institutes operate at Tehran, Isfahan, and Kerman and others are to be established at Hamadan, Ray, and Fasa. West Germany has a similar agreement for the setting up of vocational schools. About sixty agricultural schools are in operation.

The urgent need for more and more teachers is met by over seventy teachers training colleges, and about fifty teachers training institutes. Several of these colleges train teachers who live with tribal groups, conducting migratory schools.

Private schools have steadily increased in number. Some cater primarily to the children of foreign residents, preparing them to enter colleges in their home countries. Typical of this category are the American International School at Tehran, the Tehran American School, and the Iran Zamin School at Tehran. Others fill the felt needs of Persian families to send their children to "good" schools, additional attractions being the prestige attached to such attendance and the opportunity of obtaining a sound grounding in a foreign language. Private schools are supervised by the Ministry of Education, which issues permits and sees that each curriculum corresponds to those of the government schools. It also assigns government teachers to these schools: the lower the tuition charged by a school the higher the proportion of these teachers.

The demand for learning English outstrips all available facilities. The schools offer four years of English in the lower grades and three in the secondary schools, with ninety-three percent of these students choosing English as the preferred second language. Both the Iran-America Society and the British Council conduct English-language teaching programs in their centers at Tehran, Mashhad, Isfahan, and Shiraz, while in Tehran there are several highly rated private institutions teaching English, and countless smaller ones. The acquisition of English is not limited to the student level. Many adults attend one of the programs just named, and the industrial and business establishments have their own special English classes. Almost everyone in Tehran seems to understand a little English, and families of the higher social levels speak it fluently.

Early in the twentieth century wealthy families began to send their children abroad for higher education. Beginning in 1926 about one hundred students a year were selected to study in Europe and the United States at government expense. After World War II this trickle of official students was swelled by a flood of those going abroad on private means. In 1972 there were some 21,000 students abroad. Well over forty percent were in the United States, over 4,000 in West Germany, and between 1,000 and 2,000 in England, France, Austria, and Turkey. Under 3 percent are on government scholarships, and only about 7 percent are women. In contrast to the very large proportion of students in the Iranian universities in the fields of social sciences and the humanities, those abroad study medicine, business management, physics, mathematics, engineering, electronics, and other technical subjects. They are supervised by the Students' Affairs Office of the Minister of Science and Higher Education, and by officers in Iranian embassies and consulates. Many of the students marry foreigners and a considerable number become disoriented with respect to Iran and remain abroad. Anti-regime and anti-imperialist propaganda is carried on by a very small minority of these students, who also engage in picketing,

street demonstrations, and even bombings. The Confederation of Iranian Students Abroad has been active in many countries, enjoying a large membership and holding annual conferences. Acting on the belief that this organization was
controlled by communists and other revolutionary elements,
in March 1971 the government declared that membership in
it was an offense under the law providing for punishment for
those who act against the security of the country or of its
independence, and called on its present members to resign.

The most recent figures on literacy in Iran come from the
1978 figures. At that time the literates were 50.8 percent of
the entire population of six years of age and over, with only
22 percent of the females literate. However, in the urban
centers the literacy figure was 68.3 percent. Adult education
courses were first undertaken during the reign of Reza Shah.
In 1965 a National Committee for the Campaign Against
Illiteracy was established in association with the UNESCO
sponsored International Institute for Adult Literacy Methods. Adults are taught to read and write in classes held in the
cities, by the members of the Literacy Corps in the villages,
and in several pilot programs in selected regions.

Iran's education system is constantly under review and discussion. A conference held at Ramsar in 1968, attended by
the Shah, issued a so-called charter of Iran's educational revolution as it applied to higher education. Among its forty-three
clauses were those calling for closer contacts between teacher
and students, freedom of student expression, and student
management of their affairs. Over the last several years
strikes and demonstrations have occurred at the universities,
in particular at the University of Tehran which has been
closed by the authorities for limited periods. Some of the protests have concerned the curriculum and others related to
examinations. The authorities seek to make a distinction between those students with legitimate grievances which deserve
consideration, and those, allegedly inspired from outside, who
attempt to arouse their fellows with such slogans as, "the
white revolution is a lie, there should be a red revolution."
Faculty members of the University of Tehran publish articles
critical of the present system. One wrote: "Our present gen-

eration of students has realized that reading and memorizing the dull scholastic materials in the manner that we teach them is no longer useful, and that many new and exciting materials are now plentifully available." Another stated: "Today the incursion of Western culture is so extensive that it has to a great degree robbed Eastern societies of their creative abilities and is imposing its special patterns of life in all fields. We appeal to young men and women to wake up to the perils of the present trend and to realize that without the foundation of a national culture, the trappings of Western culture can only lead to the destruction of the country."

A subject that is raised periodically is the advisability of replacing the present Arabic alphabet and script with the Latin alphabet. Advocates point out that the Arabic alphabet is not as suited to Persian as is the Latin one, and this is perfectly correct, as well as the fact that the Latin alphabet is more easily mastered. Its opponents insist that such a change would mean that much of the body of Persian literature would be lost, because of the impossible job of putting it all into new publications in the Latin script.

Public Health

Throughout the entire country widespread disease and illness are related to undernourishment and to low levels of housing and living. However, the programs of social and economic reforms include increasing concern for the problems of public health.

Infant mortality is high in both towns and villages: 160 out of every 1,000 babies die before the age of twelve months. Cholera is rarely found in Iran. Outbreaks of typhus have occurred in recent years, but on each occasion strict quarantine measures have controlled them. A virulent type of smallpox was once a real menace, but the massive vaccination of school children and adults has all but eradicated the disease. Venereal diseases are all too common, although the disease types seem to be less destructive than those in western countries.

Trachoma is widespread, as is the *salak*, a festering boil apparently caused by the bite of a sandfly, which often leaves a considerable scar.

Water- and food-borne diseases, such as typhoid fever and amoebic dysentery, are prevalent but are less damaging than might be expected, since the people seem to have developed a partial immunity against recurrent attacks.

Tuberculosis strikes both farmers and townspeople. Only recently has malaria been recognized as the cause of the fevers which sapped the strength of a large percentage of the population. Though malaria was known to be rampant along the damp coastline of the Caspian Sea it had not been realized that the malaria mosquitoes also infested the high altitudes and dry climate of the Iranian plateau. Then, over a number of years, programs of spraying houses and mosquito breeding areas with DDT has all but eliminated this scourge.

Water-borne diseases and those transmitted through human waste are prevalent in Iran, although the hot, dry climate does act as a neutralizing agent to cut down on these types of infection. Few towns in Iran have sewage systems. Towns and larger villages employ a primitive septic tank arrangement with the disposal shafts frequently located under the streets and lanes. In the smaller villages both private and communal privies are found and it is common practice to collect the human excreta to fertilize the field crops.

Within recent years striking progress has been made upon the installation of systems of pure, piped water. Shiraz led the way when a private benefactor, Mohammed Nemazee, supplied funds for the entire system, including a purification plant. Work began on the system for Tehran about 1950 and in 1955 the first connections were made to houses in the city. Here, as at Shiraz, there are numbers of free outlets located throughout the urban areas. Now piped water systems are completed in scores of towns.

Foreign-owned and -operated hospitals rendered a tremendous service to the people of Iran. Over a long period of years the Presbyterian Board of Foreign Missions built and ran hospitals in the north of the country and in them trained both doctors and nurses, thereby helping to eradicate the popular belief that nursing was not a very respectable occupation. The American hospitals were in operation in Tehran, Kermanshah, Hamadan, and Tabriz. The Church of England conducted medical work at Isfahan, Shiraz, and Kerman, and the USSR maintains a large hospital at Tehran.

One measure of a country's health care is statistics which indicate the total number of doctors, of hospitals, and of hospital beds. For Iran the available figures are contradictory, and hence unreliable. This situation may result, in part, from a desire to gloss over the inadequate level of medical facilities. In 1978 there were said to be 15,000 doctors in Iran, 555 hospitals, and 56,800 beds in these hospitals. These hospitals included 175 operated by the Ministry of Health, 180 private hospitals, 96 hospitals of the Red Lion and Sun Society, 24 connected with the universities, and 16 under the Social Security programs for employees of the government. Possibly the number of private hospitals has been inflated, since other sources put the number of hospitals at under 500. Currently all the hospitals of the Ministry of Health are being transferred either to the Red Lion and Sun Society, or to the medical colleges of the universities. The Red Lion and Sun Society deserves the highest commendation: its well designed hospitals have been erected in neglected places, such as Bandar Lingeh on the Persian Gulf. There is a good deal of local criticism of the level of hospital care: private hospitals are said to charge exorbitant rates, while others are characterized as being dirty and crowded.

For a good many years the practice of opium smoking helped to undermine the constitutions of many Iranians. While it offered a facile escape from deadly drudgery, its continued use sapped the energy and the will to act, and its cost absorbed so much of the addict's limited cash that from the economic point of view the practice was disastrous. About a decade ago a law was passed which completely forbade the cultivation and processing of the opium poppy. Recently this law was rescinded and cultivation permitted in limited areas under strict state control: the reason for this change was to sell the processed opium abroad through legal channels, and hence do away with illegal production and smuggling. At the same time very stringent penalties, including the death sentence, were imposed on smugglers of opium and its derivatives.

Cultural, Intellectual, and Related Activities

The educated Persians have abundant opportunities to be well informed on both local and international affairs. Several bookstores in Tehran and scores of street kiosks stock foreign newspapers, magazines, and books, and the weekly news magazines from the United States are sold in quantity. The Ministry of Information operates the Pars News Agency which subscribes to the leading international news services, and provides local and foreign news to the press and to the radio and television stations.

Persian is, of course, the principal medium for local publication and within recent decades the language itself has undergone "Persianization." In 1937 the Iranian Academy made a study of the language of the country, and found that throughout the centuries so many Arabic words had been taken into Persian that it actually contained more Arabic than Persian words. Writers had found that they could command a larger audience if they wrote in Arabic, and there was also that fact that a knowledge of Arabic was excuse for a display of erudition. Persian histories and other works of the tenth and eleventh centuries had contained only about five percent Arabic words, but by the fourteenth century the more erudite authors were using as much as ninety percent Arabic. It was also pointed out that the language was weak in technical terms and that many European words related to such subjects as mechanics, aviation, engineering, chemistry, botany, hygiene, and medicine were being taken into the language unchanged. The Academy stressed the fact that good Persian words existed for nearly all the Arabic ones in general use, and decided to meet both these problems by publishing periodically pamphlets containing lists of purely Persian words to replace the Arabic, and of new technical terms based upon appropriate Persian roots. Turkish words were also replaced.

Lists of technical terms continue to be issued at intervals. The program was never so stressed that all foreign words were "purged" from the language and the results have been salutory. In general the ornate style and elaborate imageries of earlier times have given way to directness and clarity of expression. More recently the Iranian Academy all but disappeared from the local scene, and in 1969 an Academy of the Persian Language was established which issues word lists and conducts research on phonology, grammar, and ancient and modern dialects.

Newspapers and periodicals operate under a law enacted in 1955, and amended by decree in 1963. Persons wishing to publish a newspaper must be of Iranian nationality, over thirty years of age, of known good character, and without any criminal record; they must have sufficient funds to publish for three months. According to the decree of 1963 newspapers may not continue publication if their circulation is under 3,000, and magazines of less than 5,000, with exceptions for art, literary, and scientific publications: this decree has not been enforced. The violation of a number of prohibitions, such as attacks on the royal family, printing false news, revealing military secrets, publishing material injurious to Islam, or printing obscene material, may result in suspension of publication, fines, and imprisonment. Since 1965 reporters and press photographers have been required to obtain accreditation from the Ministry of Information.

In 1972 Iran had 25 daily newspapers, with 15 of them published at Tehran, and the others at Shiraz, Tabriz, Ahwaz, Mashhad, and Isfahan. Their daily circulation is about 200,000, a very low figure in relation to the population of the country. Weekly newspapers numbered 74, with 16 of them at Tehran; their circulation is about 80,000.

Several Tehran papers have well established positions. *Ettela'at*, owned by 'Abbas Massoudi, began publication in 1927. An afternoon paper, it appears in either eight or twelve pages, normally supports the government in office, and circulates about 80,000 copies. *Keyhan*, its leading rival, is also an afternoon paper and has about as large a circulation. Both of these papers publish a small format airmail edition for subscribers abroad. Both organizations also publish English-language morning dailies: *Ettela'at* with the *Tehran Journal* and *Keyhan* with the excellent *Keyhan International*. *Ayandegan* appeared on the scene in 1967 as a morning paper to compete with the two afternoon papers named, but it has not

seriously challenged their priority in circulation. In the second rank may be mentioned *Post-Tehran*, *Mehr-i-Iran*,
Paigham-i-Emruz, *Azhang*, and *Mard-i-Mobarez*; most are
of four pages and circulations may range from 2,000 to 7,000
copies. Associated with the Iran Novin party are *Nedayi Iran
Novin* and others of the smaller dailies; with the Mardom
party, *Rah-i-Mardom* and *Mehr-i-Iran*; and with the Iranian
party, *Iranian*.

The more important dailies provide detailed coverage of
local and international events, and offer political commentaries, articles on economics, social criticism, essays, modern prose
and poetry, and a calendar of events. Advertising abounds.

There are some 240 periodicals—weeklies, monthlies, and
quarterlies. Most are published at Tehran and about 60 are
issued by the various ministries, banks, universities, social and
cultural societies, and other semi-official organizations.

In the realm of the perodicals the Ettela'at publishing
house, which has its own excellent printing facilities, puts out
a weekly, *Ettela'at Haftegi*, a monthly, *Ettela'at Maianeh*,
and a women's monthly *Ettela'at-i-Banuan*. Its rival organization has a children's weekly, *Keyhan-i-Bachcheh-ha*. *Zan-i-
Ruz*, Today's Woman, is a weekly with a circulation of about
80,000. These magazines and several others, such as *Tehran
Mosavvar*, *Sepid-o-Siah*, and *Javanan* feature rather sensa-
tional material, including stories and novels, glamour and sex
articles, and confidential material on the lives of local and
foreign stage and screen stars. *Zan-i-Ruz* offers a matrimonial
service, with prospective clients matched by computer. A good
proportion of the material is lifted from foreign periodicals.
A weekly, *Khandaniha*, founded in 1940 after the pattern of
the *Reader's Digest*, has attained a circulation of about 30,000.
Reprinting provocative material from foreign and local papers
and magazines and including material critical of local conditions and events, it plays an important role on the local scene.

Periodicals with more serious intellectual pretensions include *Rowshan Fekr* and *Ferdowsi*. Satirical magazines
include the very popular weekly, *Towfiq*, and *Karikatur*, published by Iran's leading cartoonist. Literary journals include
Yaghma, *Armaghan*, and *Sokhan*, as well as *Rahnama-yi*

Kitab, a monthly which contains scholarly book reviews and original articles relating to literature. There are also periodicals in the fields of medicine, commerce, education, architecture and art, the petroleum industry, science, agriculture, astrology, cinema, and many others. All the periodicals are listed in *A Directory of Iranian Periodicals*, published at Tehran in 1969 in English by the Iranian Documentation Center.

The state of book publishing in Iran has given rise to considerable debate as to why the market is so small, or, in a few words, why so few Persians read books. As in some other fields, statistics from different sources on the number of books published annually are contradictory. It would appear that between 1965 and 1973 the number of books published each year rose from 1,000 to about 2,000, but this latter figure includes text books and translations of books from foreign languages. The average printing is 2,000 copies, which is too small a number to be economically profitable to the publishers and writers. Recently the introduction of small format paperbacks is resulting in larger sales. Small format is stressed, since, just as in France, many books are published in both hard cover and paper cover. Several investigations have been held into the cause of this "book crisis" which is portrayed as an obstacle to the cultural and intellectual development of the society.

A copyright law was not passed until 1970, and this law offers protection only to local writers. Scores of foreign books are pirated each year by photographic methods, and protests by foreign governments, publishers, and writers to the government of Iran have been unrewarding.

In addition to the regular commercial publishers, there are several very active organizations. The Institute for Translation and Publication, a branch of the Pahlavi Foundation, has published over 300 titles which include translations from foreign languages and some original works. It is not above pirating foreign books. Franklin Book Programs, a non-profit organization with its main offices in New York City, has put out several hundred translations of American books in cooperation with their original publishers. These include novels, but the emphasis has been upon books relating to intellectual, social, and scientific fields. Franklin has also arranged for the writing, design, and printing of millions of school text books for Iran and Afghanistan. The University of Tehran operates a very busy press, publishing well over one thousand titles, primarily books written by members of the university on their special subjects. Other organizations, such as the Society for National Monuments, also publish. The Institute for the Intellectual Development of Children and Young Adults has an admirable program: it publishes original children's stories with a Persian background which are illustrated in color by some of the country's most talented artists. The Society of the Friends of the Book issues an occasional publication, specializing in the reproduction of calligraphy.

"New" literature came into prominence in the 1920s and 1930s as the result of Western influence on the short story and the novel, forms which had not been developed in Iran. Writers of that period whose works are still read include 'Ali Dashti, Muhammad Hejazi, and Muhammad 'Ali Jamalzadeh—some of the latter's writings have been translated into European languages. Sadiq Hedayat, who died tragically in 1951, continues to enjoy great popularity for his short stories, written in colloquial language and combining psychological insight with sharp satire. His writings, also available in translation, appear to have influenced more recent writers, such as Sadeq Chubak, who writes of the lives of common people, employing the stream of consciousness technique. Jalal Al-i-Ahmad's novels, among them *The Curse of the Soil*, explore village life, with its assorted characters and local conflicts, in depth. In 1961 something of a literary bombshell burst at Tehran with the publication of a 900-page novel, *Ahu Khanom's Husband*, by an unknown writer 'Ali Muhammad Afghani: its sales exceeded those of any other recent publication. The story of a baker who, already married and with a family, takes a second wife, it explores the complexity of this human triangle, in the local idiom. Other writers look into the lives of butchers, rice farmers, and truck drivers, in contrast to the earlier novelists who wrote of the follies of the upper classes.

While prose was coming under Western influence, the poets of these same years continued to follow traditional forms and conventional imagery. They did, however, take up new subject matter, and sang of political freedoms and the need for social change. Given the priority of poetry over prose as noted earlier, it will be apparent that there were too many poets of merit to mention them by name. A sharp break with tradition was heralded by the work of Nima Yushij (1897-1960) who, in effect, threw out classical poetry. He altered the formal rhyming schemes, and expressed his visual imagery, drawn from his surroundings, in symbolistic forms. Some poets of more recent date have discarded rhyme, rhythm, and meter in favor of free verse. The latest expression of this approach is reflected in the work of a group which calls itself the New Wave: social and political subjects are totally discarded in favor of the free association of thoughts and images.

As for the less formal aspects of literature there is great interest in the collection and publication of folk poetry and folk tales, with some of this material featured on radio and television.

The advent of the cinema and television struck a serious blow to the legitimate theaters of Iran, and today there are only thirteen in the country, of which number seven are in Tehran. The theater does survive in southern Tehran and in the provinces. In a traditional form, the play takes place on a bare stage. The theme is stated, the actors come on in their turns and improvise upon the theme. The characters may vary, but virtue is always triumphant. There may be a young hero, or an older man, the *Hajji-Aqa*, of less honorable intentions; the *khanom*, a lady heroine who is young; a female servant; and the *kaku siah*, a bumbling black servant whose comic efforts do much to resolve the action.

More formal Persian drama was long dominated by the passion plays. Secular productions came to the stage as translations of European plays, and then local writers moved into the field of social satire and in the reign of Reza Shah went in for patriotic themes and moral guidance. The recent renaissance of play writing seems to reflect two trends: one, that of psychological insight into the relationships of villagers and city dwellers, and the other, the direct influence of the avant-garde foreign playwrights. The restricted number of legitimate performances is greatly augmented by the many performances given on radio and television.

Tehran is very fortunate in having its new Rudaki Hall, a most elegant opera house. Operated under the Ministry of Arts and Culture, it offers a very active program of events. It has its resident orchestra whose concerts are augmented by those of visiting orchestras. Operas and ballets, both foreign and locally composed, are produced, and folk dance troupes drawn from several regions of Iran perform.

Initiated in 1967, the Shiraz Art Festival is held every year in the autumn. Performances are given at Shiraz and Persepolis, and include orchestras and solo musicians and singers, plays, and dance recitals. Renowned international soloists and groups are invited, and a striking "Sound and Light" program is offered at Persepolis, set against its towering ruins. At the end of October the Ministry of Arts and Culture holds an Art Festival at Tehran and in many towns; its programs include amateur theatrical groups, presentations of the traditional theater as described earlier, and musical ensembles.

There are some 400 movie houses in Iran, with 120 of them at Tehran. At Tehran and elsewhere the newer houses are very spacious and are air-conditioned. The cinema is extremely popular, so popular that at Tehran there is a black market for tickets to certain films. Tickets are rather inexpensive, usually 40 cents or less. Foreign films comprise the bulk of the screenings, about 200 a year. Half of these films are American, followed by films from Italy, France, England, and India: they are all dubbed in Persian at Tehran. The less elegant houses show many Grade B Westerns, both because of their popularity and because of their lower rental costs.

The Iranian film industry got off to a very late start, with the first full length film produced after World War II. Currently there are about ten major studios at Tehran, and some 70 films are turned out each year. Some have been highly profitable, but the commercial films come under criticism for their lack of sophistication and an inclination to copy foreign productions. All films are screened prior to showing by the

Cinema Censor Board. Persian films are featured in two annual film festivals held at Tehran; one sponsored by the Ministry of Arts and Culture and the other by the cinema owners and film magazines. Prizes are awarded. The Institute for the Intellectual Development of Children and Young Adults sponsors the annual Tehran International Festival of Films for Children.

Recent years have seen a notable increase in musicology and the performance of music. Students of the history of Persian music now emphasize that its origins are to be found in the Sasanian period and have tried to destroy the theory that it was merely an offshoot of the music of the Arabs. Composers, trained in Western musical theory, have sought to combine traditional songs and themes into melodic sequences and these symphonies and concertos have been performed by the Tehran Symphony Orchestra and other ensembles. Popular music, especially "movie" music is increasingly preferred over the traditional melodies which many Persians feel are sad and not in touch with the times. Composers and singers are kept very busy with a number of new records issued each month, and about 12,000 a day purchased in Tehran alone. Tape recordings are also produced locally. Until protection was extended to singers and composers under the recent copyright law, some firms took their material directly from radio and television performances onto tapes and records.

Folk dance groups brighten many occasions. With the men and women clad in their colorful regional costumes, the dancers are accompanied by musicians playing the *reng*, the traditional accompaniment to these dances. The musicians also perform as ensembles: their instruments include the *setar*, a four-string lute, the *tar*, similar to the setar but with six strings, and the *satur*, a kind of zither or dulcimer. The drums are played with both hands; they are necked pottery vessels with skin stretched tightly across the open lower ends.

Modern Persian art has, of course, been influenced by international styles and fads, but it usually retains a distinctive reflection of the strong artistic traditions of earlier Iran. While abstract painting has its vogue, it seems fair to say that much of the better painting is in the field of semi-abstraction. Figures of girls, so stylized as to resemble bright cutout figures, are overlaid with patches of calligraphy—calligraphy having long been a highly esteemed art in Iran. Other paintings stylize and recompose the figures and forms of classical miniature painting. Landscapes are not naturalistic, although their stylization may fall short of semi-abstraction: some of these paintings are done in the muted tones of the parched fields and hills, others display disassembled forms whose bright colors are overshadowed by huge blazing suns. Many of the recognized painters have studied at the Tehran College of Decorative Arts, others have received training abroad. Their works are on display in a number of art galleries in Tehran—and not for bargain prices.

Sculpture has not the established tradition of the other arts and production remains somewhat limited. Fortunately, or unfortunately, the modern sculptors copy the international fashion of creating birds, beasts, and unrecognizable forms from scrap metal.

It may be said that interest in all the arts and culture of Iran is on the increase. The Persians have become "tourists" in their own land, visiting important historical sites throughout the country and taking countless pictures. Instead of collecting Russian porcelain, Bohemian glass, and French furniture and paintings of the romantic school, the wealthier Persians have turned to interests in the ceramics of Ray, Sultanabad, Kashan, and Gorgan, in miniatures of the Qajar and Safavid periods, in ancient bronzes, and in the highly decorative contemporary arts and crafts. However, the sparse attendance at most of the museums of the country seems to indicate that these interests have not affected the other classes of society.

Radio Tehran came into initial operation in 1940. Now one of the units of Radio Iran, it has a number of transmitters, including one of 2,000 kilowatts and an FM multiplex transmitter. It broadcasts on medium and short waves with one of the short-wave programs that of its foreign service, the Voice of Iran. Its three local programs are tailored to different audiences. In general, about half the broadcast time is given over to foreign music, and considerably less to local music. About one-tenth of the time is devoted to lectures and talks, a somewhat larger share to news, and a relatively small amount to drama. The so-called miscellaneous time includes special programs for women, children, farmers, and sports lovers. Commercials are confined to specified time slots, with rates varying according to the broadcast time from about $150 to $175 a minute.

Over 150,000 radio sets are made or assembled annually at Tehran and at least that many are imported every year. In 1973 there were over 3 million in use. Many were small battery-powered transistors: youthful shepherds tending flocks carry these transistors, as do children strolling through the streets of the towns.

Iran's first television station, TVI, came on the air at Tehran in 1958 as the result of a five-year monopoly awarded to a privately owned company. The same company opened a station at Abadan in 1960. When the monopoly came to an end, the company was permitted to continue its operations. The Abadan station serves a relay station at Ahwaz which sends the programs on to the region of the Persian Gulf, including much of the Arabian littoral. A station operated by American army personnel assigned as advisors to the armed forces of Iran opened in 1960: its range is limited to the capital and its vicinity, and its programs come canned from the United States. In 1964 the Ministry of Education inaugurated its own closed-circuit television station.

National Television, NTV, came on the air in 1967. Since its equipment employed the French linear system rather than the American one of TVI, previous owners of sets had to fit them with converters in order to receive its programs. There are National Television stations in seventeen other cities. They receive the NTV programs via microwaves through a large number of relay stations, and some of them also produce their own programs.

While NTV and TVI offer rather similar programming, NTV does place more emphasis upon educating and informing the public. TV films, features, panel games, and scientific programs from foreign sources are dubbed in Persian, while locally produced material covers this same range of subjects plus programs for women and children, interviews, music, drama, and the news. The average viewer is said to spend 40 hours a week in front of the set.

Television sets are assembled in eleven plants at Tehran at a current rate of 120,000 sets a year. Fairly recently the importation of sets was prohibited as a means of protecting the local industry, but this restriction appears to have been lifted in favor of a customs tax of 20 percent *ad valorem*. In the absence of reliable statistics, it may be estimated that there are 600,000 sets in Iran. Advertising rates are based on a sliding scale according to the number of minutes a month which are purchased. For ten minutes a month each minute costs $500, while for 120 minutes a month the price falls to $300 a minute.

There are an increasing number of youth organizations. The National Scouting Organization was established during the reign of Reza Shah. For a number of years after World War II it was relatively inactive, but now it is in a flourishing condition. There are nearly 150,000 boys scouts and an unstated number of girl scouts who enjoy some 390 scout camps. In the last few years a large number of *khaneh javanan*, youth houses, have been established throughout the country.

In international athletic competitions the Persians have been outstanding in wrestling and weight lifting. The most popular sport is soccer, followed by basketball, volleyball, and pingpong. Women take part in all these sports except the first two named. Skiing attracts enthusiasts to the slopes and lifts to the north and west of Tehran. Under the Physical Culture Organization there is a Sports Federation for each of those named, as well as for a number of others.

A huge sports complex just to the west of Tehran includes a 100,000 seat stadium, enclosed structures for swimming and other sports, and a very large artificial lake; it is the site of local and international competitions.

Secular charitable organizations are comparatively new to Iran as formerly those with means followed the principles of Islam in creating religious endowments or in distributing charity according to the system of voluntary taxation. The

Red Lion and Sun Society, established in 1942, has an annual budget of about $15 million. In addition to its hospitals, it runs clinics, nursing schools, nurseries, and vocational schools, and has its own youth organization. The Imperial Organization for Social Services, established in 1947, is very active in the fields of health and education; its principal source of income is the nation-wide sale of lottery tickets. The Pahlavi Foundation was established in 1961, and has a capital of $130 million. Its multiple activities include the fields of health, education, agriculture, the publication of books, the management of its hotels, and investment in industry. The Farah Pahlavi Charity Organization operates orphanages in Tehran and the provinces. The Institute for the Protection of Mothers and Children was established in 1940 and derives its income from taxes on cinema tickets and gasoline sales, and from the national lottery. The National Association for the Protection of Children works in cooperation with CARE to provide hot lunches for 600,000 school children. The total number of these organizations is twenty-four.

Changing Social Patterns

It would appear to this writer that in modern Iran social patterns, attitudes, and behavior seem to have gone through two overlapping periods of change, the second one more directly reflecting influences from the West.

In this first period, which can be assumed to have begun early in this century, Iran was said by the Persians themselves to be dominated by one thousand families. In fact, there appear to have been about 150 families which possessed great influence and prestige, with approximately half this number on a somewhat higher level than the others. These families intermarried to an extraordinary degree. They included landlords who lived at Tehran rather than in the provincial towns near their major holdings. They comprised more than a third of the members of the Majlis and two-thirds of the membership of successive cabinets, and some of them rose to the highest ranks in the armed forces. They engaged in interlocking commercial enterprises, frequently with self-made individuals whose energy and acumen enabled them to marry into this social elite. These families had the means to provide higher education, usually abroad, for sons and relatives, and it was these "doctors," that is, doctors of philosophy, who played important roles in economic and social planning at somewhat later dates.

Mutual interests among the elite were promoted through the medium of the *dowreh*, or circle. Its members met at intervals to discuss these interests, as well as intellectual subjects and politics, and individuals might be members of several circles. In addition, individuals not of the elite came into these circles, often because their Western education had brought them into association with its members.

While it is reasonable to suppose that these families were devoted to the preservation of the *status quo*, this did not prove to be the case. Many members were active supporters of Dr. Mossadeq and his National Front, and many more were in the forefront of the land reform and other programs of social and economic change.

This period was marked by the rapid emergence of an urban middle class. At Tehran its numbers grew by leaps and bounds to include industrialists, entrepreneurs, professional men, educators, writers, shopkeepers, bankers, managers, contractors, and those engaged in the service trades. The bustle and urgency of business pervaded the urban atmosphere, and the urge toward material accomplishments and the acquisition of possessions became increasingly strong. With the accelerated tempo of life came a practical interest in efficiency, precision, and the avoidance of delay, with none of these features rooted in the older social patterns. With national progress dependent on the adoption of the techniques, methods, and outlook of the Western nations, social attitudes had to change.

In attempting to establish basic motivations of societies, some sociologists and anthropologists have put forward the theory that behavior in one type of society is determined by the urge for expiation of guilt, while in another type the emphasis is on feelings of shame as the motive for individual conformity to its customs. It may be suggested that in Iran there was a shift from the guilt reaction to that of shame.

However, shame was more negative than positive. Among the better educated there was shame at their lack of patriotism, and shame in their questioning of allegiance to traditional ethics and to the primacy of the extended family within society.

The second period began in 1941 after the abdication of Reza Shah when an era of "new freedom" produced a rash of critical commentaries, among which the hostile Marxist evaluations of the Iranian political and social orders were the most knowledgeable and severe. In addition, socialists, conservatives, and religious figures presented their own views of society, as did literally hundreds of ephemeral newspapers at Tehran. There was a full decade of this general disorientation. The Persians have always been marked individualists, and this aspect of Persian social behavior became enhanced in private and public life. Dr. Mossadeq himself remarked: "In our Parliament each deputy has his own personal opinion and this is why bills, even the simplest, cannot be passed rapidly by this body." This exaggerated individualism found expression in negativism, expressed in the form of destructive criticism of authority and of the country's institutions, groups, classes, and individuals. On the political scene negativism found formal expression in Dr. Mossadeq's policy of negative equilibrium.

The Tudeh party was active in attempting to create and foster dissension, distrust and hostility between social classes, between the people and officials, between teachers and students, and between landlords and peasants. The traditional feelings of tolerance seem to have been replaced by unrestrained expression and open hostility.

The first phase of this second period began to come to an end in August 1953 with the public choice of the established institutions of the country over the alternative choice, that of continuing turmoil and social strife. The end of the first period was foreshadowed by the Land Reform program and by the White Revolution of 1963. It would be possible to say that there was a third period from 1953 until the present, but actually it is the shift in public attitudes in these years, together with the decline in the role of the leading families, that tie the second phase of the second period to its first.

In these more recent years there have been increasing signs of social integration and of more cooperative social behavior. Social mobility is on the rise, promoted by education, and evidenced by the presence in cabinets, other high positions, and in the Majlis of many individuals from "unknown" families. The stress on individualism is losing out to participation in clubs, societies, institutes, and organizations devoted to the public welfare. Negativism is on the wane, primarily because Iran's remarkable economic progress no longer inclines the Persians to feel that other nations are superior to their own. Individuals in responsible positions find a sense of pride and gratification in their personal contributions to social and economic programs. The so-called "brain drain" of somewhat earlier years according to which qualified Persians educated abroad failed to return to Iran has eased off as positions commensurate to their training and talents are readily available. As stated earlier, social and economic programs bore the imprint of Western social patterns.

Of course, a fully integrated society remains a distant, probably an unobtainable, goal, and too optimistic a picture of the changing social patterns is not intended. There are many who deplore the all-pervading influences of the West, believing that they are destroying the nation's cultural traditions and the cement of its society, and placing the urban Persians in a state of limbo between this ancient culture and a foreign one which has no roots in Iran. "Guidance" by the government of the press and during the electoral processes, together with the lack of full public freedoms serves to deaden the voices of constructive critics. Thus, the opposition is limited to elements, within the country and abroad, who engage in destructive activities. This is not to say that the press is uncritical of conditions and events within Iran, but the subjects so developed are ones on which the government is urged to take corrective action. The problem of leisure time is much in the fore. As the society becomes more affluent, more such time is available to workers and their families, and the feeling is that most of this time is wasted. This feeling stems from the belief, stated earlier, that man's goal was to move toward identification with the divine, rather than to engage in self-indulgence. It also ties in with discomfort over the fact that the better educated pay only lip service to the principles and obligations of Islam.

VI. THE GOVERNMENT

Constitutional Government

THE Constitution of 1906 has been modified by a supplement of 1907 and amended by Constituent Assemblies in 1925, 1949, 1957, and 1967, while sections which were outlined in general terms have been made both more specific and more comprehensive by legislation. The document is modern and liberal on the pattern of the constitutions of the European democracies, but also contains provisions relating to the state religion of Iran—the Shiʻa sect of Islam. It provides for a government composed of three branches: the executive, whose power is vested in the cabinet and in the government officials who act in the name of the Shah; the judicial, composed of a hierarchy of courts from district courts up through a Supreme Court; and the legislative, whose bills do not become law until signed by the Shah.

The legislative branch, or Parliament, consists of the *Majlis*, or Assembly, and the Senate. The Majlis has powers superior to those of the Senate, notably in financial matters and foreign affairs.

Up to the nineteenth session of the Majlis, inaugurated in 1956, some 136 members were elected for terms of two years. Then the term was increased to four years and the number of deputies was periodically increased proportionate to the rise in population to provide one deputy for every 100,000 people. Thus, the twenty-third Majlis, inaugurated in 1971, had 268 members. In this session the Iran Novin Party was represented by 225 deputies, the Mardom Party by 36, and the Iranian Party by only one. The Armenian, Jewish, Assyro-Chaldean, and Zoroastrian communities each chose one deputy to represent their special interests.

Until fairly recent years the composition of the Majlis was fairly consistent: large landlords were in the majority, followed in numbers by merchants and religious figures. Currently the number of landlords has greatly declined, and there is an increasing number of government employees, professional men, and a scattering of workers and farmers.

[229]

Although the Senate was provided for in the Constitution, it was not brought into legal being until 1949 and first met in 1950. It numbers 60 members: 30 are appointed by the Shah, 15 elected from Tehran and 15 from the rest of the country. Its role is to act, if necessary, as a conservative check upon the more exuberant Majlis. The Senate is composed of rather elderly individuals among whom are former prime ministers, cabinet ministers, retired generals, and distinguished business and professional men. In the Sixth Senate, inaugurated in 1971, 28 of the elective seats were won by the Iran Novin Party and two by the Mardom Party.

Parliament receives proposed legislation from the cabinet ministers and usually refers it to one of eight committees, then receives the committees' reports and discusses the bills, which may be passed by a majority vote. Deputies may also initiate bills if the bills are signed by fifteen other deputies. Parliament may ask ministers of the cabinet to appear and answer oral or written questions, and normally one or more cabinet ministers attend each session.

The Constitution gives Parliament the authority to control the policies and acts of the cabinet on matters vital to the interest of the nation. It must approve any disposal of State property or funds, internally or foreign managed concessions and monopolies, government borrowing, and proposed construction of roads and railways. The Shah may order the Majlis dissolved during a session, and must give his reason for such action. He may also withhold his signatures from bills passed by the Parliament and return them for reconsideration. Sessions of the Majlis are held in the Baharestan, set in a large park in the heart of Tehran, while the Senate meets in its ornate new building. The proceedings of the public meetings are published in an official journal. Private or secret sessions may be held upon the written request of ten deputies or of one of the cabinet ministers.

Earlier election laws have been replaced by one enacted in 1963. It provides for universal suffrage for citizens of twenty years of age and older, with the exception of those with criminal records and those on active duty with the armed forces, police, and gendarmerie. The nation-wide elections are held on a single day, in July or August, and the voters present their identity cards at the polling places. Candidates who might be "unsuitable" members of the Majlis appear to be eliminated from consideration by the method in which the Iran Novin and the Mardom parties select their slates of candidates.

Political Activity

Since 1942 many political parties have been founded, waxed and waned. The first to be formed, some of which were sponsored by merchants and landowners and used Tehran newspapers for their party organs, were weak in organization and vague of program. The parties of this first period bore names of such popular appeal as Justice, People, Fatherland, Fellow-Comrades—the last offering a socialistic program, while other rather liberal parties included the National and the Iran.

During the war years the most active and influential parties were the *Eradeyi Melli*, or National Will, and the *Tudeh*, or Masses.

The National Will party was led by Sayyid Zia al-Din Tabatabai, who had been for a short time in 1921 Prime Minister of Iran and had then been exiled from the country for many years. In 1943 he returned to Iran and was elected to Parliament. His party published a detailed program proposing social and economic reforms, but he and his followers were soon attacked as reactionaries by the Tudeh party, while he himself was singled out for the hostile attentions of the Russian press and radio. The party had followers in the Parliament and published newspapers at Tehran; its funds were said to derive from local businessmen and it was believed that its leader had British backing. After his arrest in 1946 the leader and party faded from the political scene.

The Tudeh party was organized in 1942 by a group of educated men most of whom had been imprisoned under Reza Shah as suspected Communists. It strove for a tight organization, was exceedingly vocal, solicited members from the newly formed trade unions, and elected eight members to the 1944-1946 session of Parliament. In the province of Azerbaijan the militant branch of the party assumed the name of

the Democratic party of Azerbaijan and led a separatist movement which isolated that region from the rest of the country
from the end of 1945 until the winter of 1946. A detailed
consideration of the history of the Tudeh party and of Communism and subversion in Iran appears under a following
heading.

Among the parties founded in 1942 only the Iran party
continued active throughout the postwar years. In 1944 it
entered into temporary alliance with the Tudeh party, an
action which has been held against it by recent governments.
The Iran party issued pamphlets describing its liberal program but the intention was not to attract a mass following.
Instead, the effort was concentrated upon winning over key
figures within the government and in private life to become
unpublicized but active proponents of its policies.

About 1950 Dr. Mozaffar Baghai founded his Toilers'
party, designed to speak for the industrial workers of Iran.
His lieutenant, Khalil Maleki, previously a Tudeh party
leader, broke away to establish the Third Force, a socialistic,
anti-communist grouping of intellectuals. Lunatic fringe
groups also appeared: the national socialist Sumka party,
mystically dedicated to the active life, to courage and discipline, and to the struggle against colonization, communism,
capitalism and the parliamentary system; and the Pan-Iranists, ardent nationalists who are resolved to recover territories
lost in the last century to the Russian empire and to other
neighbors of Iran.

In 1950 the National Front, a loose amalgamation of individuals and groups, attached itself to the personality and
prestige of Dr. Muhammad Mossadeq. In time it came to
include the Iran party, the Toilers' party, religious leaders,
and independent deputies in the Majlis. Members of the Iran
party became cabinet ministers and ambassadors. By the
middle of 1953 the Sumka party, the Pan-Iranists, and the
Toilers' party had deserted Dr. Mossadeq but after his fall
his most vehement nationalist followers gathered into the
underground National Resistance Movement which put out
a clandestine paper, The Path of Mossadeq, in 1956 and 1957.

In July 1960 the National Front was reactivated in order

to contest the elections for the XXth Majlis: it was composed of the Iran party, the Pan-Iran party, elements of the Third Force, the National Resistance Movement, and the tiny People of Iran party. Although the elections of August 1960 returned none of its members, in January 1961 its head, Allahyar Saleh, was elected from Kashan.

After the summer of 1961 the National Front concentrated its efforts upon demanding that Parliamentary elections be held, and in strengthening its organizational structure. The National Front had a Central Council of up to 50 members and a small Executive Committee, headed by Allahyar Saleh of the Iran party. While its headquarters were at Tehran, it had considerable strength among students and self-exiles in Europe and in 1962 opened a number of provincial branches in Iran. In 1963 the National Front held mass meetings and demonstrations, and published anti-government leaflets. In 1965 the arrest of several of the leaders led to a disintegration of the group. Its activities continue abroad with the publication of a paper, *Bakhtar Imruz*, directed at Iranian students.

The *Melliyun*, or Nationalists, party was founded in 1957 by Dr. Manuchehr Eqbal, then prime minister, to be the majority party in the Majlis. In the same year the *Mardom*, or Peoples, party was established by Asadullah Alam. In 1963 the *Iran Novin*, Modern Iran, party displaced the Melliyun party. At first headed by Hasan Ali Mansur, then prime minister, since January 1965 it has been led by prime minister Amir 'Abbas Hoveida. Both these parties strive to enlist broad public participation, holding congresses and party celebrations, and maintaining scores of branches throughout the country. The Mardom party represents the loyal opposition in the Parliament, actively criticizing internal policies and actions of the government, but solidly supporting its foreign policies.

The Iranian party was founded in January 1971. Led by a former head of the Pan-Iran party, its aims are nationalistic and include hostility towards Communism.

For a number of years consideration had been given to broadening the system whereby the administration of the country was centralized at Tehran. Initial steps to increase local responsibility were made far-reaching by a law of June

1970. Under its provisions councils are now elected for each of the 14 *ostans*, or provinces, each of the 448 *bakhs*, or districts, and each of the 141 *shahrestans*, or urban centers. These elections are contested by the political parties named above and by independent candidates who have been successful in winning about 20 percent of the total number of seats.

Political Subversion

Political subversion, in the form of Communist activity in Iran, has a relatively long history.

It was long the aim of Russia to acquire all of Iran down to its warm-water ports. The testament of Peter the Great expressed the goal in these words, "And in the decadence of Persia, penetrate as far as the Persian Gulf"; while in 1941, Molotov stated to his German counterpart in Moscow, "The center of gravity of the aspirations of the Soviet Union is the area south of Batum and Baku in the general direction of the Persian Gulf."

The Communist Party of Iran, which held its first meeting at Enzeli (Pahlavi) in June 1920, was the outgrowth of the Communist *Edelat*, or Justice, party which had been founded in 1918 among the Iranians working in the oil fields at Baku. From its inception, it appears to have been headed by A. Sultanzadeh, for a time head of the Near Eastern Section of the Commissariat for Foreign Affairs of the USSR. In September of the same year he and Haidar Khan represented the Communist Party of Iran at the Congress of the Peoples of the East at Baku. The new party, aiming at the overthrow of the Shah, operated from the Caspian littoral of Iran, but the re-establishment of central authority in this region in 1921 by troops directed by Minister of War Reza Khan deprived it of a secure base, and it went underground in Azerbaijan and Khorasan. Sultanzadeh, a prolific writer, recounted its early successes and subsequent trials and indicated that its most effective members went to Tehran to support and direct the workers' unions that had been established in 1920.

This trade-union activity included the publication of newspapers at Tehran. Many of the individuals active at this time were to reappear on the local scene many years later as lead-

ers of the Tudeh party or of the Democratic party of Azerbaijan, and a few examples should be of interest. Jafar Kavian,
who had been active since 1918, was to emerge as minister
of the people's army of Azerbaijan in 1945. Jafar Pishevari,
who put out the paper *Haqiqat* at Tehran, became the prime
minister of the Azerbaijan regime. Reza Rusta had established peasant unions in Gilan before being sent to Tehran;
years later he headed the Tudeh party unions.

Members and Communist leaders of the Tehran unions
celebrated May Day with public meetings until 1929, when
Reza Shah ordered the self-styled "freedom lovers" arrested.
Those named and a number of others remained in jail or in
enforced residence until the fall of 1941.

While those of its members engaged in trade-union activity
had consistently denied any Bolshevik ties, the Communist
Party of Iran was not as reluctant to disclose its plans and
intentions. In 1926, both the party and the Young Communist League of Iran urged the Soviet Union to intervene mili-
tarily in Khorasan where a localized revolt had broken out.
In 1927, the Communist parties of Turkey and Iran held a
congress at Urmia (Rezaieh), in 1928 Sultanzadeh and
Shareqi represented the Communist Party of Iran at the Sixth
Congress of the Comintern, and at the end of 1929, Sultanzadeh was still the head of the party.

Then, in 1930, a series of articles by a Soviet defector,
George Agabekov, was published in Paris. At the final stage
of his career he had been resident general of the GPU in Iran
from 1927 until 1929; his articles described in fascinating
detail the secrets of Soviet espionage and subversive activities
in Iran and included the names of many Iranian agents of the
Soviet Union. As an initial reaction to his revelations, some
350 minor Soviet agents and activists were rounded up and
in 1931 another 100 or so arrested, including such an important figure as 'Abbas Mirza Eskandari, a prince of the former
Qajar dynasty.

The Majlis passed a comprehensive bill designed to rem-
edy the fact that existing criminal legislation provided no
penalties for individuals plotting to overthrow the government by force, and in 1932, thirty-two persons charged with

espionage on behalf of a foreign power were brought to trial and all but five convicted. At this same time the Iranian government expressed concern over the publication in Germany of pro-Soviet Persian-language newspapers. One, the *Setarehyi Sorkh*, or Red Star, the organ of the Communist Center of Iran, appeared at Leipzig, and another, *Paykar*, or War, published at Berlin, was succeeded by *Nezhat*, or Insurrection. In response to official protests, the German government closed down these papers. The fact that thirty years later Leipzig and Berlin once again became centers of Communist activity directed against the government of Iran is not without interest.

The Communist Party of Iran was heard from again in 1932 in a long article published in *International Press Correspondence*, the organ of the Comintern. The article discussed the position of the party with respect to Reza Khan [sic] and the dispute between the government of Iran and the Anglo-Iranian Oil Company, and propagated the slogans of the party overthrow of the monarchy and establishment of a workers' and peasants' government in Iran. The party continued to send Iranian Communists trained in Russia into the country. One of them was 'Abd es-Samad Kambakhsh, who was arrested in 1933 and released in 1941, just in time to become a founding member of the Tudeh party. Ardashir Ovanessian was educated at the Young Communist School at Moscow, was sent back to Iran in 1933, and was arrested in 1934; after 1942 he was a member of the Central Committee of the Tudeh party and later director-general of propaganda for the Azerbaijan regime.

After 1932, the Communist Party of Iran vanishes from the scene. The stern authority of Reza Shah had made ideological operations within Iran impossible; the center in Germany had been closed down; and the USSR was, apparently, unprepared to permit the party to operate openly on its territory. Since the disappearance of the party took place so long ago, what relation has this fact to the contemporary scene in Iran? First, it brings up the question of what has become of the Communist Party of Iran. Once such a party has been organized does the Soviet Union ever permit it to be dissolved? At any rate, the professional revolutionaries were followed by a group of amateurs who became active about 1930. They were Marxists, but had not been trained in the Soviet Union. These individuals grouped themselves around Dr. Taqi Erani, who had gone to Germany for advanced studies in 1922 and become deeply interested in Marxism. At Berlin he came into contact with 'Ali Bozorg Alavi, Dr. Morteza Yazdi, and others. In 1930 he and his friends returned to Iran, where he began to build up a personal following of students and professional men, and in 1933 he started publication of a monthly intellectual magazine called *Dunya*, or World.

In May 1937 Dr. Erani and fifty-two of his associates were arrested; forty-nine of them were brought to trial in November 1938. Forty-five were convicted of being members of a Communist party; receiving funds from a foreign power; and using such funds for propaganda purposes. They were sentenced to prison terms of up to fifteen years. They were welcomed to the Tehran model prison by the real professionals and "the old prisoners were grieved, but happy at the same time." These are the words of Jafar Pishevari, who went on to say, "They were all well educated, but they had not had our experience. Their appearance gave us an opportunity to teach them. They learned from us how to resist and endure, and we gave them moral support. Doctors and professors who were the intellectuals of Iran acted like trained political warriors." Dr. Erani died in prison in 1940, but the others were released in September 1941.

On January 30, 1942, the *Tudeh*, or Masses, party was created by the followers of Dr. Erani, with a small leavening of the professional Communists, who looked to three members of the Eskandari family, 'Abbas Mirza, Sulayman Mohsen, and Iraj, to add an air of respectability to the new party. Former prisoners Dr. Reza Radmanesh, Dr. Morteza Yazdi, Dr. Feridun Keshavarz, 'Ali Bozorg Alavi, and Dr. Muhammad Bahrami, the new party's Marxist theoretician, became members of the Central Committee of the Tudeh party of Iran.

The Tudeh party of Iran avoided any mention of Communism, insisted that it had no ties with the Soviet Union, and stated that it was an antifascist party working for the defense of the constitution of Iran and for the independence of the country. In the first two years of its very active life, attention was concentrated upon organization and upon recruitment of members. Ardashir Ovanessian, a member of the party's Central Committee, wrote a book entitled *Fundamentals of the Organization of a Party*, which explained in detail the Soviet concept of democratic centralism, or the control of the many by the few. He described the basic party unit of the *huzeh*, or cell, and explained the duties of the member of a cell and his relationship to the higher organisms of the party.

By 1944, the Tudeh party had gathered most of the local trade-unions into its Central United Council of Trade Unions, headed by Reza Rusta, and in 1945, this body affiliated with the World Federation of Trade Unions.

Party support came primarily from the industrial workers of the country and from the intellectual class. Open direction was by the intellectual elite represented by the former associates of Dr. Erani, but the real power was wielded by the professional Communists who remained in the background and even, such as in the case of Pishevari, denied that they had any connections with the Tudeh party. In 1944, eight of the Tudeh party leaders were elected to the Majlis; and because of their discipline and singleness of purpose managed to come close to dominating debate in that body. Most of them were elected from the northern provinces of Azerbaijan, Mazanderan, Gilan, and Khurasan where Soviet troops were in occupation and where the Tudeh party was more powerful and influential than the officials of the central government.

In 1945, the Tudeh party steadily grew in strength and influence, and the government seemed unable to counter what was finally recognized to be a serious threat to the political stability of the country. However, the clear indications that the international conflict would soon end with the defeat of Germany caused the Soviet Union to take steps to be certain that the withdrawal of its protection would not result in the repression of its adherents in Iran. What better way to associate this necessity with its positive goals than to establish a permanent foothold in the country?

On September 2, 1945, the Tudeh party in the northwest was suddenly replaced by the Democratic party of Azerbaijan although the leaders of the party were not all in favor of this decision. Groups of armed members successfully attacked army garrisons and gendarmerie posts, while, in November, the commanders of Soviet forces refused to permit government reinforcements to move toward the areas in open revolt against the central authority.

All the Soviet instruments were now in place as the culmination of a carefully worked-out plan. About 1935, many thousands of Iranians long resident in the USSR had been repatriated: hundreds had received training in the Soviet Union and had been told to await instructions. When the Soviet troops entered northern Iran in 1941, these individuals flocked to welcome them, and pointed out the influential individuals loyal to the Iranian government who were then removed from the occupation zones. After September 1945 they formed the nucleus of the armed forces of the Democratic party of Azerbaijan, as well as supplying a major part of the membership of the so-called legislative bodies of the autonomous regime.

Other persons emerged as leaders of the new regime. One element was made up of Iranian Communists who had fled to the USSR to escape arrest during the reign of Reza Shah and had come back in 1941 in the uniforms of officers and men of the Soviet occupation forces. Typical of this element was Faraj Dehqan, who appeared in 1941 as a Soviet officer at Astara and in 1945 was the commander of forces of the Democratic party of Azerbaijan in the Ahar District. Another element was composed of long-time Communists, most of whom had deliberately avoided identification with the Tudeh party of Iran. In addition to individuals named earlier who were prominent in the Azerbaijan regime, such as Pishevari, Kavian, and Ovanessian, there were many others. Dr. Salamullah Javid, minister of the interior, had been a party to the attempt made in 1920 to separate Azerbaijan from Iran and had then fled to the USSR. Returning to Tabriz in 1929, he had been arrested for Communist activity and jailed until 1941. Muhammad Beeria, the barely literate minister of education, had received training in Russia and been sent to engage in trade-union activity at Tehran; he had been arrested in 1931.

In the hands of these individuals and their Russian advisors, the Azerbaijan regime had emerged as a totalitarian state. Its provincial National Assembly proclaimed the Autonomous Republic of Azerbaijan, approved a constitution, entrusted all authority to Pishevari and his associates, and then disbanded. On the southern fringes of Azerbaijan, Soviet officers brought the Kurdish Republic of Mahabad into being in December 1945. In April 1946 the "national governments" of these two areas concluded a treaty which stressed their fraternal relations and their determination to resist the central government. Both regimes faced increasing difficulties after the withdrawal of the Soviet occupation forces in May 1946 and both collapsed when Iranian troops moved into the region in December of that year. At Tabriz many of the "democrats" were slaughtered by the populace, but others, including Pishevari, Beeria, Ovanessian, and Gholam Yahya Daneshiyan, commander of the forces of the regime, managed to escape and cross the frontier and make their way to Baku.

The loss of Azerbaijan was a bitter blow to the Tudeh party as well as to the USSR itself. After December the party theoreticians met, as is customary in Communist parties, to discuss why the movement in Azerbaijan had failed, what errors the party itself had committed, and what changes in the party line seemed indicated. The fact was that the USSR had pushed its local instruments and organizations too hard, misjudging the assumed receptivity of the people of Iran to "liberation" from its so-called reactionary regime.

While still engaged in the reorganization and regrouping of its forces, the Tudeh party suffered a more serious blow by being outlawed in February 1949 following an attempted assassination of the Shah. Most of the members of its central committee managed to flee abroad, including ten who escaped from detention at Tehran at the end of 1950. Early in 1950, a second echelon of party members was engaged in the clandestine printing and distribution of leaflets attacking the government in terms similar to those employed by the Soviets at this time. In 1952, Reza Radmanesh discarded the fiction that the party was a national, liberal one with no foreign ties when he informed the Nineteenth Congress of the Communist Party of the Soviet Union, "The Tudeh party is the sole party of the workers of Iran; inspired by the ideology of Marxism and Lenin . . . [it] is based upon democracy and socialism, at the head of which are to be found the U.S.S.R. and Josef Stalin, our great standard bearer."

Suddenly, the rise of Dr. Mossadeq offered the party a new lease on life. The xenophobic, anti-British nationalism of the Mossadeq regime was most welcome to the Soviet Union, which ordered the Tudeh party to exploit the situation and direct part of its effort to discrediting the United States. Scrawled messages reading "Yankee Go Home" appeared on the walls of Tehran. Numerous front groups were quickly formed, including the Peace Partisans, Democratic Youth, National Society Against Colonialism, Democratic Women of Iran, Society Against Illiteracy, and Society to Defend the Rights of Villagers, and scores of newspapers were published by these ephemeral organizations. In addition, the Tudeh party newspapers, *Mardum*, *Razm*, and *Zafar*, were clandestinely printed and circulated at Tehran, party cells sought fresh recruits, and arms were collected and stored.

Although the Tudeh party was officially opposed to the Mossadeq Government, it directed its major effort to helping it stay in power. In so acting it followed a familiar Soviet tactic of encouraging a bourgeois nationalist movement, attempting in due course to form a common front with the movement, and planning to take over from that movement should the opportunity present itself. Thus, on July 21, 1952, the party took part in a mass demonstration in favor of Dr. Mossadeq, and a year later celebrated the anniversary of the occasion with a huge rally.

The party had been able to emerge into the open following a decision in March of that year by a Tehran court that members of the Tudeh party were not liable to prosecution for activity against the constitutional monarchy and in favor of Communist doctrines: it is difficult to understand why the government of Dr. Mossadeq allowed this ruling to go unchallenged.

During the events of August 16-19, 1953, the Tudeh party attempted to exploit a period of confusion and uncertainty to swing popular opinion in favor of a republic. Whether the second-level party leaders who were in charge at Tehran believed that the party was ready for such a test is not known; at any rate, they followed Soviet instructions. Groups of party members went far beyond expressions hostile to the Shah which were uttered by members of the National Front: statues of the ruler and of his father were pulled down from their pedestals, showcases were smashed to remove pictures of the Shah, and inspired crowds were encouraged to surge through the streets shouting in favor of a republic. However, the effort failed. On the one hand, popular sentiment was not prepared to go all-out against the monarchy, and on the other, the party could not put enough people in the streets to be able to resist counterattacks on the part of the National Front itself, and, later, of the armed forces. After the removal of Mossadeq, the Tudeh party issued a pamphlet admitting errors and blaming traitors within the party for its failure to take command of the situation in August 1953.

In September 1954 the so-called Organization of Tudeh Officers came to light, and within a few days some 600 officers and noncommissioned officers of the Iranian armed forces were under arrest. As the interrogations proceeded, it was revealed that the atmosphere of the Mossadeq period had been most favorable to Tudeh party activity within the armed forces, since the organization had grown in size from some 100 members to over 600 persons. Following military trials, a score of officers were executed and many others sentenced to long prison terms. Within a year, the government published a *Black Book* that described the Organization of Tudeh Officers. The text traced connections of the group with the USSR and listed confessions of espionage activity, of attacks on the constitutional monarchy, and of training of Tudeh youth groups for street fighting and guerrilla warfare against the government. In addition, the organization was to assassinate Dr. Mossadeq after he had been successful in eliminating the monarchy from Iran and then stage an armed Communist take-over of the government.

The shock of these disclosures spurred the Parliament to pass a bill directed against organizations that promoted Communism, that attacked Islam and the constitutional monarchy, and that contained members of the Tudeh party or other groups declared to be illegal.

Leaders of the Tudeh party who had fled from Iran established the headquarters of its Executive Committee at Leipzig, broadcasting its attacks on the government of Iran from Radio Peyk-i-Iran. Dr. Radmanesh informed the Bulgarian Communist Party that he brought greetings from the Iranian masses, at the head of whom was the Tudeh party. The winds of chance favored Dr. Radmanesh and his associates. Following the July 1958 revolt at Baghdad, the Communist Party of Iraq, long underground, was permitted to operate openly and the Qassem regime was on the best of terms with the USSR. In 1959, Dr. Radmanesh, Reza Rusta, Iraj Eskandari, Dr. Feridun Keshavarz and a number of other Iranian Communists arrived in Baghdad. One aspect of their mission was probably to put their experience in organizational activity and propaganda at the service of the Communist Party of Iraq, another was to take advantage of the fact that Iraq was the only country of the Middle East in which Communist activities were not illegal, and a third was to establish direct relations with remnants of the Tudeh party in Khuzistan, immediately adjacent to southern Iraq. Couriers went from Baghdad not only to that area but also to Fars, Kurdistan, and Azerbaijan in Iran.

At the Sixth Congress of the Tudeh party of Iran held in Leipzig in September 1959 there was discussion whether the party should move its headquarters to a Middle East center of Communism at Baghdad. Steps were also taken to form a union with the Democratic Party of Azerbaijan, and Gholam Yahya Daneshiyan was welcomed as a new member of the Executive Committee of the Tudeh party. However, in Iraq the Communists had overestimated their strength, and their criticism of the Qassem regime brought retaliatory action which limited their freedom of speech and action. As a result, Radmanesh and his little group were forced to move back to Leipzig and the Seventh Congress, held in the summer of 1960, seemed to have closed on a pessimistic note, according to the account supplied to the Communist press.

Near the end of October 1961 Dr. Reza Radmanesh had the opportunity of addressing the Twenty-second Congress of the Communist Party of the Soviet Union and spoke these words:

> The present rulers of Iran, headed by the Shah, who took over the country through a *coup d'état* against the legal government of Dr. Mossadeq, have turned our country into open territory for invasion and plunder by the imperialists. . . . The defeat of the usurper regime is now near, and the struggle of our masses during the past two years shows that we are now at the beginning of the rise of a great national liberation movement. . . . Our party . . . is more than ever gaining strength and strives to unite all the democratic and national forces in a united front of all the masses against imperialism and feudalism, in order to struggle in union for the overthrow of the usurper regime, the establishment of democratic freedoms, withdrawal from aggressive blocs, and the establishment of a national government.

His remarks echoed the then current Soviet line of encouraging national, anti-imperialist movements in areas where local Communist parties were too weak to act effectively on their own. It represented a call for the return of Dr. Mossadeq to power in the expectation that given a parallel situation in Iran the Tudeh party would not repeat the errors of 1953 and be more successful in exploiting such a national, anti-imperialist government.

In 1962 the Central Committee of the Tudeh party appeared to have four major functions: maintaining party discipline, preparing propaganda material, conducting recruiting activities among the thousands of Iranian students in Europe, and keeping open channels of communication with party members in Iran. Of these functions, only that of its propaganda work comes to public attention. The party published *Mar-*

dum, a four-page Persian-language monthly, in Europe and mailed copies throughout Europe, to Iran, and to the United States. Its material was voiced in Persian and Turki by party members and announcers over Communist-controlled radio stations at East Berlin, Leipzig, Sofia, Moscow, and Peking— the peripatetic Reza Rusta has been heard from Peking. In addition, material communicated by party members in Iran was used on the programs of the so-called National Voice of Iran, a clandestine station located in the vicinity of Baku, possibly at Bazuna.

After the assassination of Qassem and the installation of a new revolutionary regime in Iraq, Tudeh party activity was renewed in Baghdad. The party cooperated with General Timur Bakhtiar, a former head of SAVAK, who had left Iran in 1962 and turned up in Baghdad in 1969 to engage in a so-called Iran National Liberation Movement. Bakhtiar was murdered in 1970, and in 1971 Radmanesh was expelled from the Central Committee of the Tudeh party for grave deviations from the party line, including his association with Bakhtiar. Iraj Eskandiari was named Secretary-General of the party in his place. In 1973 the party line towards Iran was carried by Radio Iran Courier, the apparent successor of Radio Peyk-i-Iran. The party was deserted by a number of its extremist members. Some formed the Revolutionary Organization of the Tudeh party, and others established *Tufan*, or Storm, the Marxist-Leninist Organization of Iran which assumed a pro-Chinese posture: both organizations engaged in sabotage in Iran.

The hostility between Iran and Iraq, discussed in a later section, found one outlet in the subversive political activities sponsored by the Iraqi regime. It created the National Front for the Liberation of Arabistan, which later became the Popular Front for the Liberation of Ahwaz: the goal was to detach the province of Khuzistan from Iran, and the methods included sabotage and propaganda. The regime set up a Baluchistan Liberation Front and cooperated with the Popular Front for the Liberation of Oman and the Arabian Gulf, sending saboteurs from both these organizations into Iran. It also trained other saboteurs, including Kurdish tribesmen, who were sent into northwestern Iran. In 1972 Baghdad put the Voice of Iranian Freedom on the Air, and in 1973 this station was joined by the Voice of the United Front of Nationalities in Iran.

The various elements named above engaged in guerrilla activity, attacks on police posts, bank robbery, numerous bombings at Tehran and other points in Iran, the murder of individuals including an Iranian general and an American officer on duty in Iran, attempted abductions, and the hijacking of airplanes. Between 1971 and 1974 about 75 of these individuals were apprehended and executed, a like number killed in gun battles with police, and several hundred more arrested and given long prison sentences.

The Shah

Muhammad Reza Pahlavi was born in 1919. His titles are *Shahinshah*, or King of Kings—Iran is officially an empire—and *Aryamehr*, or Light of the Aryans. His empress, Farah, is addressed as *Shahbanu*, the Shah's Lady. Married in 1959, their children include Crown Prince Reza, born in 1960, Princess Farahnaz, born in 1963, 'Ali Reza, born in 1966, and Princess Leila, born in 1970.

The Shah is a tireless exponent of programs of social reform and economic progress, and a severe critic of local administrative shortcomings. An exceptionally intelligent and shrewd person, his constant contacts with officials and private persons keeps him so abreast of developments that he can speak convincingly and in great detail on any subject. Officials are shifted from post to post and new ones named as he continues to search for a more effective administration of the nation's affairs. Carrying out his Constitutional role as head of the armed forces, he has become remarkably well informed on military matters. Aircraft are a special interest, and he personally pilots all types of planes.

His wide-ranging views about Iran and international affairs are expressed on his travels, in speeches, press conferences and interviews, and in writing. Since 1953 he has made official and private visits to some forty-five countries, always making himself available to the local press. And he tours his own country at frequent intervals. At home he holds frequent press conferences, and grants audiences to visiting representatives of other nations, businessmen and scholars. Each year he delivers a No Ruz message to the nation, reviewing the progress of the year just ended and the plans for the new one. He addresses the opening meetings of each session of the Parliament, speaks on numerous public occasions, and releases his instructions to the cabinet ministers. His published writings, available in English, include *Mission for My Country*, *The White Revolution*, and collections of his speeches. Writings, speeches, and interviews all reflect a firm confidence in the bright future of Iran as it joins the ranks of the highly industrialized nations.

The Shah and other members of the royal family are served by the Ministry of the Imperial Court. The minister has equal rank with other members of the cabinet, but does not attend cabinet meetings. The ministry handles the royal finances and has divisions concerned with ceremony and protocol, with the upkeep of the royal properties, and correspondence and records. The palaces built by the Pahlavi dynasty in Tehran are in two locations, one in the western quarter of the capital, and the other in the Shemiran region. Until recently the family would spend the winters in town and the warm season in the Shemiran, but the town palaces are now to be converted to serve as museums and for other purposes.

The Imperial Inspectorate was established in 1958 to investigate complaints of citizens against government officials and departments. Abolished in 1962, it was revived in 1968 under a new law which gave it more comprehensive powers. Each year at least a score of investigating teams travel throughout the country to hear, and to take action, on such complaints.

Government Employees

The first civil service code was enacted in 1922, and the current one in 1966, with amendments of 1972. Employees fall into two categories: permanent and contract. Monthly salaries are according to twelve grades, with fifteen steps within each grade. The grade at which an employee enters the service is determined by his education and special qualifications, and for every two years of service he advances one step. The basic salaries are adjusted to cost of living figures, and there are several types of allowances, the totals of which may not exceed the basic salary. Grade 1, Step 1 receives about $60 a month; Grade 5, Step 1 nearly $100; Grade 9, Step 15 about $400; and Grade 12, Step 15 nearly $600.

According to the State Organization for Administration and Employees Affairs there are some 350,000 government employees. At least 50,000 of the lower grade employees are illiterate. Of the remaining 300,000, about 50 percent are high school graduates, over 10 percent college graduates, and more than 5,000 employees have Ph.D. degrees. About 25 percent of the employees are women, largely engaged in teaching. Employees benefit from a health insurance plan, and may retire after twenty years of service.

As does any bureaucracy, that of Iran has its shortcomings. The salaries of the lower grades are so relatively low in relation to the cost of living that many employees moonlight or engage in corrupt practices. Locally the bureaucracy is criticized for inefficiency, immersion in petty details, and indifference to the public whom they should serve. For example, "In the lower echelons there are many government employees who would like to have hordes of people refer to their desks in order to make others believe that they may be important, and, in particular, prove their worth to the higher bureaucrats. This is why these employees, by resorting to some devious tricks and sending their clients back and forth, manage to have each person refer to their desks, say, three times instead of once."

The Prime Minister and the Council of Ministers

The Prime Minister is appointed by the Shah. He selects his cabinet ministers and presents them for royal approval and then presents the program of his government to the Parliament for its consent. The cabinet resigns prior to each new Parliament and then the Prime Minister presents the same members or a different combination to this body. He may also make cabinet changes according to his wishes. The Prime Minister remains in power until he chooses to resign or until the Parliament records a majority vote of "no confidence." The Prime Minister longest in office has been Amir 'Abbas Hoveida who was named in January 1965 and was still in his post throughout 1973.

Cabinet Ministers must be Persian citizens of the Muslim faith and may not be ranking princes of the royal family. Parliament may submit to the Supreme Court charges of misdemeanors against cabinet ministers.

Continuity within the executive branches of the government is not seriously dislocated by the recurrent changes of Cabinets. Each ministry is headed by an Under-Secretary and the major divisions of the Ministry by Directors-General, high officials who usually retain their posts for a number of years and are thoroughly familiar with the normal activities of their ministries. Then too, the various cabinet posts tend to be held by a limited number of individuals so that an experienced official will be familiar with the operations of several ministries.

Each cabinet usually includes one or more Ministers without Portfolio, or Ministers of State. Their function is to advise and assist the Prime Minister and to perform special duties, including important investigations and inspections. In recent years there has been a considerable increase in the number of ministries, and there continue to be changes in the total number and in the names of some of them. In 1973 they were as follows: Finance, Foreign Affairs, Interior, Labor, Economy, Defense, Culture and Arts, Education, Science and Higher Education, Roads, Justice, Water and Power, Housing and Development, Posts, Telegraphs and Telephones, Information, Health, Agriculture and Natural Resources, and Cooperatives and Rural Affairs. Rather than describing the functions of each ministry, certain topics relative to a number of them will be discussed.

Foreign Affairs

The Ministry of Foreign Affairs maintains embassies in forty-six countries, and has ambassadors to the United Nations and to the Vatican. Iran was a founding member of the United Nations, and belongs to all of its specialized agencies. According to the Sa'adabad Pact, signed by Iran, Afghanistan, Iraq, and Turkey at Tehran in 1937, the signatories agree to refrain from interference in each other's internal affairs, to respect their mutual frontiers, to refrain from aggression against each other, and to consult together in the event that an international conflict should threaten their common interests: this document remains inoperative.

Iran is a member of the Central Treaty Organization (CENTO), the successor to the Baghdad Pact, along with Turkey, Pakistan, and Great Britain. The United States is not a full member of the organization but is represented on several of its committees. CENTO has shifted its concern from mutual defense against Communist aggression and subversion to the development of transport facilities and telecommunications between Turkey, Iran, and Pakistan.

The Regional Cooperation for Development (RCD) was founded in 1964 by Iran, Turkey, and Pakistan. Its Secretariat is at Tehran, and its several committees deal with the subjects of industry, petroleum and petrochemicals, trade, transport and commerce, public administration and technical cooperation, and social affairs. Typical of its efforts are decisions as to which member states should be the site of selected industrial enterprises.

Long plagued by British and Russian interference in its internal affairs and embittered by the Anglo-Soviet invasion of August 1941, Iran welcomed the appearance of the United States on the local scene. This began late in 1942 with the arrival of troops of the Persian Gulf Command to operate the supply route through the country to the USSR, was marked by the presence of President Roosevelt at the Tehran Conference in 1943, and was heightened in 1946 by its strong diplomatic support of Iran's decision to send its armed forces to recover Azerbaijan. Then, from 1943 until early in 1945 the Millspaugh Mission was active in studying and realigning the finances of Iran.

In May 1950 Iran adhered to the United States Mutual Defense Assistance Agreement. This agreement provided for the establishment of a Military Assistance Advisory Group to Iran (MAAG). This mission joined others already functioning in Iran: United States Military Mission with the Imperial Iranian Army (ARMISH), and United States Military Mission with the Imperial Iranian Gendarmerie (GENMISH). By the end of fiscal year 1961 Iran had received from the United States some $1.3 billion, recorded under the general headings of a Mutual Security Program and a Non-Mutual Security Program for economic aid, military assistance, and other grants and loans.

In October 1950 the United States concluded with Iran the first Point Four agreement to be reached with any country, and in 1951 a mission of the International Cooperation Administration began activity. Later USAID, its annual budgets mounted, with very substantial amounts going to help keep the Iranian economy afloat during the years of dispute and uncertainty following the nationalization of the oil industry. With the rapidly improving financial position of Iran American aid was terminated in 1969 and the procurement of military supplies and equipment shifted from grants to outright purchases by Iran.

Along with its decreased dependence on the United States, Iran moved gradually to a more neutral posture in its foreign policy. This shift was inspired in part by a desire to broaden its foreign relations and in part by efforts of the USSR to attract its good will. The Soviet approach was not caused by Communist political expediency, but by a desire for closer economic relations. Joint projects are mentioned in a following section, while here it may be noted that the two nations concluded a fifteen-year economic agreement in 1970 which provides for trade exchanges amounting to $1 billion in the first five years, a transit agreement in 1971, a technical and scientific agreement in 1971, and a cultural agreement in 1972. Iran is, however, troubled by the massive military aid given by the USSR to Iraq, and, on a lesser scale, by the continued Soviet support of the Tudeh party.

While relations with such Western powers as Great Britain, France, and the Federal Republic of Germany continue to be very good, there has been an upsurge in relations with the Socialist countries, largely through the initiative of Iran.

Thus, trade and economic agreements were concluded in 1971 and later between Iran and Poland, Bulgaria, Romania, Yugoslavia, and the German Democratic Republic. A significant reflection of Iran's rapid industrialization is that the countries named were to receive not only oil but also durable goods, such as buses, trucks, tires, refrigerators, shoes, textiles, and knitwear. In return Iran receives credits, commodities, and industrial equipment.

In the region of the Far East Iran has long had good relations with Japan. In 1971 Iran recognized the Peoples Republic of China, and in 1973 the countries concluded a five-year trade pact calling for annual exchanges in amounts in excess of $70 million.

Within the general area of the Middle East relations are always good with Pakistan, Afghanistan, and Turkey. In 1972 Iran and Afghanistan finally concluded a long-discussed accord for sharing the water of the Helmand river which flows within Afghanistan for nearly all of its course. Relations with the Arab states range from very good to lukewarm to hostile. The Shah and King Faisal of Saudi Arabia hold very similar views on the need to oppose the policies of the radical Arab regimes in the region of the Persian Gulf. Some years ago Iran was seized with what its officials called the Green Problem, a euphemism for the expansionist designs of Arab nationalism and Pan-Arabism. It saw these designs reflected in the efforts by Egypt and Iraq to rename the Gulf the Arabian Gulf and in subversive activity, as detailed earlier. In 1960 Egypt broke off diplomatic relations with Iran in belated recognition of the fact that Iran had extended *de facto* recognition to Israel in 1950: relations were resumed in 1970.

In January 1968 the British government announced that it would withdraw all its military forces from the Persian Gulf by the end of 1971. Following this announcement, the Shah outlined Iran's policy for the Gulf, stating that his country was an island of stability in a sea of chaos and upheaval and was best fitted to exercise leadership in the region. Later he offered to enter into any kind of a regional treaty or pact which would guarantee the area's security and has insisted that no outside power will be permitted to fill the so-called vacuum caused by the British withdrawal. Iran's current program of building up its military strength along and adjacent to the Gulf is a direct reflection of his determination to insure its security. On 30 November 1971 Iranian forces occupied the Persian Gulf islands of Abu Musa, Greater Tunb, and Lesser Tunb, the latter two near the Strait of Hormuz. With regard to Abu Musa, a prior agreement had been reached with a rival claimant, the ruler of Sharjah, while the ruler of Ras al-Khaimah denounced the occupation of the Tunbs over which he claimed sovereignty. The most violent reaction to this action came from Iraq which severed diplomatic relations with Iran.

Relations with Bahrain are cordial, those with Qatar are correct, those with Kuwait vary from good to fair according to the winds of political change, and those with the Union of Arab Emirates—Abu Dhabi, Dubai, Sharjah, Ajman, Umm al-Qaiwain, Ras al-Khaimah, and Fujaira—are good. Agreements delineating the limits of the continental shelf of the Gulf have been concluded between Iran and several states of the Arabian littoral.

Relations with Iraq have steadily deteriorated since the 1958 revolution at Baghdad. The major source of contention concerns the Shatt al-'Arab, the confluence of the combined estuaries of the Tigris and Euphrates rivers of Iraq and the Karun river of Iran. Through its length of 56 miles, until it emerges into the Persian Gulf, it is bounded by Iraq along the west bank and by Iran along the east bank. Early treaties between the Ottoman empire and Persia awarded Turkey control of the Shatt al-'Arab despite the established principle that the boundary of a waterway between two states is the *thalweg*, or mid-channel, of a river or other waterway. Thus, the boundary between Iraq and Iran was fixed at the low water line on the eastern, or Persian, shore. In 1937 Iran and Iraq concluded a frontier treaty. With reference to the Shatt al-'Arab, it was to be open to the merchant ships of all nations, with the boundary at the low-watermark on the eastern shore except for a stretch along the island of Abadan belonging to

Iran where it was the *thalweg*. Long drawn out efforts by Iran to resolve problems within this treaty failed to produce results. In February 1969 Iran declared the treaty of 1937 null and void because of Iraq's persistent violations of its terms. Iraq replied by stating it would use force to maintain its control, but backed down in the face of Iranian naval superiority. It did show its anger by expelling many thousands of Persians long resident in Iraq, and renewed this activity after the occupation by Iran of the three islands in the Persian Gulf: at least 60,000 Persians were moved across the frontier into Iran.

In addition to the subversive activity of Iraq against Iran, described earlier, the Popular Front for the Liberation of Oman and the Arabian Gulf spreads violent propaganda against Iran and most of the Arab states along the Persian Gulf.

Armed Forces and Other Organizations Concerned with National and Internal Security

The most recent law providing for compulsory military service was enacted in 1971. National military conscription is for 25 years: 2 years of active military service, 6 years of stand-by military service, 8 years of first-stage reserve, and 9 years of second-stage reserve. Deferments are granted to students and those in other categories and exemptions are provided, as well as provisions for qualified individuals to join the Revolutionary Corps.

Official figures on the size of the armed forces are not available, but the number may be between 225,000 and 250,000. The First Army has its headquarters at Kermanshah, the Second Army at Tehran, and the Third Army at Shiraz. Its components include at least 3 infantry divisions, 3 armored divisions, 4 independent infantry brigades, artillery and anti-aircraft regiments, and auxiliary units. Infantry weapons come largely from the United States, tanks from the United States and the United Kingdom, and armored personnel carriers from the United States and the USSR. With the Shah's expressed concern for the security of the Persian Gulf, forces have been shifted from areas adjacent to the USSR to south-western and southern Iran, notably to new bases at Bandar 'Abbas and Chah Bahar.

The strength of the navy is over 10,000 men. Its vessels include 5 destroyers, 4 fast frigates, 4 corvettes, 6 minesweepers, 16 patrol boats, 12 hovercraft—the world's largest fully operational squadron of hovercraft—2 landing craft, and auxiliary vessels. Most of the above vessels are of recent vintage. The major base is at Bandar 'Abbas. The Iranian navy engages in annual CENTO exercises with elements of the British and American navies: in 1970 this exercise was staged in the Persian Gulf.

The personnel of the air force numbers about 25,000. It comprises 4 wings of 5 squadrons, with each squadron made up of 16 aircraft, of F-5 (Northrop) fighter bombers, 2 squadrons of F-4E (McDonnell) Phantoms, 2 interceptor squadrons of F-4D Phantoms, and one of F-86F Sabre fighters. Also, a reconnaissance squadron of RF-5 (Northrop), and a transport wing with 56 C-130 E/H (Lockheed) Hercules, and 12 F-27 (Holland) Friendships. The 200 or so jet combat aircraft are provided with British and American SAM missiles. There are well over 200 helicopters. Most of the aircraft are of American manufacture, and many of the Iranian pilots have received their training in the United States. Major bases are at Tehran, Hamadan, Dizful, and Bandar 'Abbas. The air force takes part in the annual CENTO exercises.

Advanced training for the officers of the armed forces conducted in High Command, Joint Staff, and High Management colleges is now given in the National Defense University, founded in 1968. Para-military organizations include the Civil Defense Organization and the National Resistance Force.

In current years the Ministry of Defense is allocated 28 percent, more or less, of budget revenues. In the year 1352 (March 1973-March 1974) the budget was 693 billion rials, equivalent to $9.2 billion. Thus, this ministry received some $2.5 billion. It is also probable that purchases of military equipment were funded from additional sources.

The police force functions in the urban areas and the gendarmerie, a force of some 25,000 men, operates in the countryside: both come under the Ministry of the Interior.

In 1957 SAVAK, an acronym of the Persian name for the National Security Organization, was established by law. Its duties include providing security protection for the ruler and members of the government, gathering material on subversive activities against the state, and carrying on counter-espionage operations. Remarkably successful in penetrating the Tudeh party and other elements, it has come under criticism for its allegedly harsh and illegal methods. SAVAK comes under the office of the Prime Minister.

National Budget

The annual budget is prepared by the Ministry of Finance from figures supplied by the various ministries and other organizations. It is reviewed by the budget committee of the Majlis and then presented to the Majlis for its approval by the Prime Minister prior to the beginning of the Iranian year. The budgets for a number of years reflect the ever rising amounts of receipts and expenditures:

Year	Billions of rials
1974/75	2,445
1973/74	693
1972/73	548.5
1971/72	481
1970/71	406
1969/70	331
1968/69	279
1967/68	217
1966/67	192
1965/66	176
1964/65	144

Each budget has two separate parts, administrative and development, with the latter somewhat less than half the total amount of the budget. Oil income now provides over 80 percent of revenues. Indirect taxes, such as customs revenues, continue to far exceed revenues from direct taxes, such as income taxes. Additional amounts come from foreign loans and internal borrowing.

The Plan Organization

The Plan Organization, *Sazman-i-Barnameh*, has semi-autonomous status under the office of the Prime Minister. In 1946 a High Plan Committee was established and instructed to prepare a development plan for the country. It engaged firms of foreign consultants. In 1949 the Majlis approved the First Seven-Year Plan and empowered a Plan Organization to implement it. Beginning in 1950 the Plan Organization was to receive all the income from the operations of the Anglo-Iranian Oil Company. The nationalization of the oil industry and the resulting loss of oil income had a damaging effect on this plan; only 16 percent of the projects were executed, most of them of the so-called impact variety.

The Second Seven-Year Plan ran from September 1955 to September 1962. Its authorization was for 70 billion rials, later increased to 84 billion, of which amount 75.2 billion, or about $1 billion, was actually spent in the fields of transport and communications, agriculture and irrigation—the construction of several large dams—and industry and mines. The oil revenue going to the Plan was cut to 80 percent of the total and then to 60 percent: this short fall in funding was met in part by long-term loans totalling $362.9 million from the IBRD, the Export-Import Bank, the Development Loan Fund, and other sources.

The Third Plan ran from September 1962 until March 1968. It differed radically from the first two in that instead of being a series of individual projects of widely varying magnitude it represented a comprehensive scheme to integrate all normal and development expenditures and programs in order to increase the gross national product (GNP) by at least 6 percent a year. Secondary objectives included the creation of an optimum number of employment opportunities, and more equal distribution of income, and the encouragement of private investment in industry and agriculture. It was to be financed mainly from oil revenues, starting from 55 percent of the total in 1962 and rising to 80 percent in 1968. Total expenditures were about $2.7 billion, allocated as follows: transport and communications, 21.4 percent; agriculture and irrigation, 21.5 percent; power and fuel, 18.7 percent; industry and mines, 13 percent; education, 9.7 percent; health, 7 percent; manpower, 4.4 percent; municipal development, 3.6 percent; and statistics, 0.7 percent. The success of this Plan was reflected in the fact that the GNP grew by more than 8 percent annually.

The Fourth Plan ran from March 1968 until March 1972. It envisaged the spending of $8.134 billion as follows: new development projects, $6.4 billion; continuing expenditure on Third Plan projects, $600 million; external debt repayment, $375 million; administrative expenses and sundry projects, $134 million. The major source of its funds were represented by 80 percent of oil revenues. During the period of the Plan the GNP was to increase by 9.4 percent annually. Major projects included a steel mill, an aluminum plant, and petrochemical plants, as well as the exploitation of the country's mineral resources. During the course of the Plan increased revenues from oil and other sources added an additional $2 billion to the original allocations. The GNP grew over 10 percent annually.

The Fifth Plan runs from March 1973 until March 1978. Allocations went from $36 billion to $70 billion. The text of the law authorizing this Plan enables the government to obtain foreign loans and credits of up to $6 billion and $5 billion from local borrowing from banks and other sources. The mining sector is expected to grow at an annual rate of 23.5 percent, and the industrial sector at 15 percent. Allocations for social welfare are five times as great as in the Fourth Plan, and those for agriculture nearly three times as large. The GNP is expected to grow at 11.4 percent annually.

Indicators of Economic Growth

Local and foreign economists have emphasized Iran's very high rate of economic growth over the past decade, considering it especially remarkable since it was accompanied by general price stabilization.

One indicator of this growth is the rise in the gross national product:

Selected Years	Gross National Product
1961/62	$ 4.4 "
1964/65	$ 5.4 "
1967/68	$ 7.3 "
1970/71	$12.2 "
1972/73	$17.3 "
1973/74	$27.3 "
1977/78	$68.0 "
1978/79	$58.5 "

Another indicator is the increase in per capita income, which has mounted during years in which the population grew at about 2.5 percent a year. The following figures may be viewed with some reserve, since different sources for this information are not always in agreement.

Year	Per Capita Income
1966/67	$ 210
1968/69	$ 220
1972/73	$ 556
1973/74	$ 810
1977/78	$2,220

Reliance on long-term foreign loans is warranted for developing countries whose domestic savings are not large enough to satisfy its investment demands. It would appear from unassembled data that by 1972 Iran had obtained $1.745 billion in long-term loans. However, this amount may be too low. Additional loans were negotiated in 1972, and, as shown above, others were anticipated during the Fifth Plan. The repayment of the interest and principal amounts of these loans already represents a very heavy drain on Iran's foreign exchange holdings. Major suppliers have been the Export-Import Bank of the United States, the International Bank for Reconstruction and Development, consortiums of foreign banks, and France, Great Britain, and the Federal Republic of Germany. In addition, the USSR and the Socialist states of Eastern Europe have extended $1.310 billion in loans, of which sum $700 million came from the USSR. These loans draw only 2.5 percent interest and are repayable in goods rather than in hard currencies.

Justice and the Courts

The Ministry of Justice is concerned with the enforcement of existing laws. Iran's judicial system, modeled after that of France, is a hierarchy of courts from the district courts on the lowest level up to the Supreme Court, which is the court of final appeal but has not the power to determine the constitutionality of current legislation. Judges try all cases except those few for which juries are called, and frequently receive permanent appointments.

Since 1925 there has been a complete revolution in the Iranian concept of law. Before that time the country was subject to a system of religious law based upon the Qoran, augmented by the interpretations of learned theologians, and enforced by the Muslim clerics. This type of law not only was poorly adapted to the conditions of the modern world, but placed too much responsibility and power in the hands of the clergy. It is surprising, however, that a series of new laws could so rapidly efface a heritage of so many centuries.

In 1925 an extensive commercial code was enacted by Parliament, followed in 1926 by a criminal code, and in 1928 by a civil code of some nine hundred articles. These three basic codes established the new legal system, while certain other laws were aimed at the "modernization" of Iran. A law of 1928 abolished the so-called "capitulations," which up to that time had given foreigners resident in Iran some immunity from arrest as well as the right of trial by their own consular representatives. Laws passed in 1931 and 1935 established the legal ages for marriage and enhanced the divorce and property rights of women. Laws of 1928 and 1935 did away with the ancient Persian costume and made western clothing and headgear compulsory, and the use of honorary titles, another holdover from the autocratic past, was abolished in 1935.

The most systematic and extensive legal changes were undertaken after 1933 by Minister of Justice Davar who was given full powers to put laws into effect and then to submit them to the Majlis for approval after they had been tried out in practice. More recent governments have paralleled this experiment for short periods.

A peculiarly Persian tradition which has managed to survive the enactment of modern codes of law is that of *bast*—the act of taking sanctuary within mosques, royal stables, telegraph offices, etc., where fugitives were safe from arrest. Political figures have sought *bast* within the Parliament grounds and at the gate of the royal palace.

The Supreme Court, with twelve branches, is not only the highest court of appeal, but has wide powers in other judicial matters. Each province has a Court of Appeal, and there are 66 provincial, 168 county, and more than 200 district courts.

The first House of Justice was established in 1963 and in 1972 some 5,500 were functioning in the villages. Each House consists of five members, elected locally for three-year terms. They hear civil cases involving amounts up to 10,000 rials, and penal cases concerned with minor rural offenses. Decisions are forwarded to the nearest district court for confirmation. The success achieved by the Houses of Justice led to the creation of Councils of Justice in cities and towns. The first was set up in 1966 and in 1972 155 were in operation. Each court has five members elected from among 30 qualified candidates. Civil cases are heard for amounts up to 20,000 rials, and in cases of petty crimes and misdemeanors the councils may sentence offenders to up to two months imprisonment. There are no fees for litigation in either the Houses or the Councils.

In general, the judicial system is not held in high esteem by the Iranians. It is felt that wealth and prestige may influence the decisions of the courts and that the courts do not provide adequate protection for abuses of individual rights. It is recognized that the courts are heavily overburdened. More than 2,600,000 new cases come up each year, and it may be a number of years before a case is finally settled. However, the establishment of the Houses of Justice and the Councils of Justice are providing a more rapid and even-handed disposal of minor cases.

In all the towns and large villages there are official registration offices staffed by the Ministry, in which all such documents as marriage certificates, birth certificates, rent contracts, and property and land deeds must be recorded. This service has been of real value in establishing complete and permanent records of land ownership, for in earlier periods the boundaries of land holdings were always very loosely defined and led to many disputes over titles.

VII. THE DEVELOPMENT OF NATURAL RESOURCES

Minerals

SURVEYS indicate that Iran possesses extensive and widely varied mineral resources. Until about fifteen years ago all mining had been done by laborious hand methods, but with the erection of modern plants for processing ores more efficient systems of extraction have been introduced.

The sub-surface riches of the country are the subject of a special Mining Law, last revised in May 1957, which divides all such resources into three categories. If materials of the first category—limestone, building stone, marble, and gypsum—are found on privately owned land the proprietors may exploit the deposits and pay the government 5 percent of the value of the materials. If the materials are in the second category—metallic minerals, solid fuels, salts, precious minerals and precious stones—the owner must obtain a license for exploration and exploitation and the state receives 4 percent of their value. The third category includes petroleum deposits and all radioactive materials, and the rights of exploitation belong exclusively to the government regardless of the ownership of the land. The government may lease such deposits to local companies or may conduct its own mining operations. Foreign companies are granted concessions for the location and exploitation of petroleum or mineral deposits only through special action by Parliament.

In 1962 the first and second categories included 820 explored mines of which about 270 had been exploited to some degree. Within the last decade over 600 licenses were issued for the exploitation of materials in the first and second categories. Coal and iron ore are of particular importance as they are vital to the development of the local steel industry.

About 50 coal mines are being worked in Iran. Confirmed deposits total nearly 300 million tons, with major deposits in the Alborz range north of Tehran, in Khorasan, and at Zarand and other sites in the general region of Kerman. Production in 1971/72 was 8,750,000 tons, and this was expected to exceed over 14 million tons in the following years. Until a few years ago the major deposits of iron ore were located to the south of Arak and at Baqf, to the east of Yazd. New major deposits have been found at Gul Gowhar, southwest of Kerman, and at Chagharat and other sites near Kerman. Coal, iron ore, and limestone are shipped to Isfahan by a railway line from Zarand, 45 miles northwest of Kerman, which passes through Baqf and Yazd.

In January 1966 Iran and the USSR signed an agreement which provided that the USSR would advance credits of $286 million at 2.5 percent interest over a twelve-year period to help finance a steel mill, a gas pipeline to the Soviet Union, and a machine tool plant. In the same year construction began at the Arya Mehr Steel Mill, situated on the banks of the Zayandeh river, some 28 miles from Isfahan. Soviet engineers and technicians supervised the construction and the installation of machinery and equipment from the USSR. As stated above, the essential ingredients for steel production came directly to the site by rail. A cement block factory in the vicinity of the plant provided the material for the construction of housing for the staff and workers, some 20,000 of whom were engaged in the project. The first blast furnace came into operation in January 1972, and current production is rated at 750,000 tons of steel a year. However, later agreements with the USSR provide for additional units to raise its capacity to 2 million tons and, finally, to 4 million tons annually. The total cost of the mill has been estimated to be $1.4 billion, a figure which includes the housing project and the exploitation of the mines.

In 1973 it was reported that a mill was to be constructed which would produce steel by direct reduction of the ingredients with natural gas. It was anticipated that this mill would be erected at Bandar 'Abbas and would obtain iron ore from the Gul Gowhar mines. A new railway line would run from the mines to Bandar 'Abbas, and both finished products and ore would be exported from Bandar 'Abbas port. The cost of this mill, estimated at $400 million, is included in the Fifth

Plan, with a part of its financing to be provided by Japanese investors.

Very limited information is at hand about the Shahriyar Steel Plant at Ahwaz which went into operation in 1972 and produces 200,000 tons of iron ingots annually, converted into 400,000 tons of structural steel, reinforcing rods, and wire, including barbed wire. Scrap iron provides the raw material for the plant.

The Ahwaz Pipe Plant, established in 1967, is owned by the National Iranian Oil Company and funded in part by the Import-Export Bank. Imported steel sheets are converted into pipes of from 18 to 48 inches in diameter: its main purpose is to supply pipes for the gas line to the USSR. With the level of present production, an important percentage of its output is exported. Adjacent to this complex is the Ahwaz Rolling Mill Plant where steel sheets are produced from imported steel and from the output of the steel mill at Isfahan. Its cost, nearly $400 million, was provided by banks, organizations, and private investors within Iran.

An aluminum smelting plant, constructed at a cost of some $53 million, went into operation in 1972. Its shares are owned by Iran, 70 percent; Reynolds Aluminum Company, 25 percent; and Pakistan, 5 percent. The raw material, bauxite, is imported from Australia. The annual production is 45,000 tons of ingots and other forms of the metal: Iran takes 15,000 tons for domestic needs, Pakistan 10,000 tons, and the balance is available for export sales.

Recent surveys have established the locations of very large deposits of copper ores. The largest appears to be that at Sarcheshmeh in the Kerman region, with at least 400 million tons of sulfurous copper ore. A state enterprise, the Kerman Copper Mining Company, is to construct a plant at a cost of between $350 and $400 million which will produce 145,000 tons annually of wire, cables, tubing and other items. Technical supervision will be supplied by the Anaconda Copper Company. Other important deposits have been found in the regions of Kerman, Yazd, and Birjand. In several cases foreign companies engaged in exploration have announced their intention to engage in joint enterprises with Iran in the exploitation of these finds. Japanese interests propose to set up a copper concentrate plant near Birjand, and a Romanian group will establish a processing and smelting plant in the southeast of Iran.

Associated lead and zinc ores are mined for export in the amount of 70,000 tons annually. With the discovery of important new deposits near Baqf, a concentration plant is under construction: it will increase the lead content to 60 percent of the ore and the zinc to 52 percent. Chromite ores, found to the north of Bandar 'Abbas, are exported through that port in the amount of about 200,000 tons annually. In 1972 it was announced that the Union Carbide Company had entered into an agreement with local investors to build a chromite smelting plant.

The mines at Anarak, well within the limits of the Dasht-i-Kavir, show signs of having been worked for long centuries, originally for the extraction of silver, but now also for nickel and lead. The known gold deposits would seem to require placer mining to put the extraction of this precious metal on a profitable basis. Uranium ores have been found at Anarak and around Yazd, Isfahan, and Hamadan. An agreement was concluded with the French Atomic Energy Commission for the further exploration and exploitation of this resource. In 1971 rich veins of gold were discovered in Khorasan. Iron oxide, antimony, and magnesite are mined on a limited scale.

Chemical salts such as the borates and sulphates of various minerals are found in rich layers, while table salt is obtained either by mining rock salt from the many salt domes or by allowing the flow of salt springs to crystalize in settling basins.

Building materials including stone, gypsum, and lime are available throughout the country, and there are quarries of marble and alabaster near Maragha, Yazd, and Shiraz. The proper earths for pottery, baked bricks, and special firebrick are common. Such precious and semi-precious stones are found as the topaz, emerald, sapphire, carnelian, and turquoise; near Nishapur are countless shafts and galleries from which the matrix embedded with fine turquoises has been extracted for many years.

Petroleum, Natural Gas, and Petrochemicals

Petroleum springs were seeping up through the rock of Iran when the earliest man appeared on the plateau. Oil flows or the escaping gas from the oil-bearing beds were ignited, probably by lightning, and became the object of wonder and worship. Much later, when the cult of Zoroaster became the official religion of the Achaemenids and truth and light were venerated, some of the temples housing an eternal flame were erected over or close to springs of oil or gas. The ruins of one such monument, that of Masjid-i-Sulayman, may still be seen in the heart of the producing oil fields. Still later, more practical use was made of the natural petroleum. A European who visited Iran nearly two hundred years ago wrote: "On that side they call Mazanderan, they found the Petroleum of Naptha. It is used in varnishing and painting, and in Physick too, for the curing of raw cold Humors. Also the meanest sort of People burn the Oyl that is made of it."

In modern times the search for oil deposits began near the end of the nineteenth century. In 1901 a British subject, William D'Arcy, heard of the possibilities, and the Iranian government granted him exclusive rights for exploration and exploitation of all of the country except the northern provinces. The concession, which included the exclusive right of building pipelines to the Persian Gulf, was to run for sixty years. D'Arcy's first companies were personally financed. In May 1908 the first free-flowing well was sunk in an area about 125 miles north of the head of the Persian Gulf. The following year the Anglo-Persian Oil Company was formed in London and took over the operation of the concession. In 1912 the British government made its first contribution to the financing of this company, and later bought control of 52 percent of the voting power of the outstanding stock.

In 1932 the Iranian government, dissatisfied with certain terms of the contract between the company and the Iranian state, suddenly canceled the concession, and in 1933 negotiated a new contract. Briefly, that contract had provided for: 1) a concession until 1993; 2) limitation of the concession to an area of 100,000 square miles; 3) loss of the exclusive right to build pipelines to the Persian Gulf; 4) payment to the Iranian government of four shillings on every ton of oil either sold for consumption within Iran or exported, and 5) an annual payment equal to 20 percent of all annual dividends of the company above the amount of 671,250 pounds sterling. All these payments must total at least 750,000 pounds each year.

In more recent years the Anglo-Iranian Oil Company had six fields in production in the hilly country north and northeast of the head of the Persian Gulf. The crude oil from these fields was pumped through several pipelines to Abadan, on the Gulf, where is situated the largest and one of the most modern refineries in the world. Since the pressure of the gas in the oil-bearing strata is just sufficient to force the oil to the surface, the company developed a special process by which the heavier oils remaining after the refining process are pumped back from Abadan to the fields and injected into the strata to build up the underground pressure.

The Anglo-Iranian Oil Company employed some 65,000 Iranians and a large British and foreign staff. In 1949 the production of the Anglo-Iranian Oil Company totaled 27,200,000 tons, and in 1950 it reached 32,259,000 tons. In 1949 royalty payments by the company to the Iranian government were listed as equal to $37,779,958. The oil company also expended an amount in pounds sterling about twice that of the royalty payments for the purchase of the Iranian currency essential to its operations in Iran.

In western Iran a subsidiary of the Anglo-Iranian Oil Company, the Khanaqin Oil Company, exploited the Naft-i-Shah fields near Qasr-i-Shirin, immediately adjacent to the frontier between Iran and Iraq. The oil is pumped over the high mountains guarding the plateau to a modern refinery in the town of Kermanshah, and from there gasoline and kerosene are distributed throughout northern Iran.

In 1937 the Iranian government granted a concession covering the oil deposits of the northeastern section of the country to the Amiranian Oil Company, a subsidiary of the Seaboard Oil Company of Delaware. Survey work was undertaken at once, but in 1938 the American company relinquished its concession. In 1944 American and British oil companies displayed an active interest in obtaining concessions in the southeastern part of the country, while the USSR pressed for the creation of a joint Irano-Soviet company to exploit the possible oil reserves of the northern provinces of Iran. During World War II Soviet occupation forces operated drilling rigs at Semnan, east of Tehran, and along the Caspian coast, and Russian geologists formed definite opinions regarding the potential fields of northern Iran. In fact, Soviet soldiers were stationed at a drilling site near Semnan for several years after the end of the war in order to assert Soviet claims to a nineteenth-century inoperative concession covering this Kavir Kurian area.

Following the nationalization of the Anglo-Iranian Oil Company in 1951—a subject discussed at some length in earlier pages—the export of oil from Iran dwindled to a few thousand tons a year. On February 1, 1954, the Anglo-Iranian Oil Company announced that it was meeting with seven other large oil companies with interests in the Middle East to discuss ways of resolving the difficulties which prevented Iranian oil from returning to the world markets, and about this same time the American government stated that should five American oil firms join a consortium they would not be liable for anti-trust prosecution. Technical experts representing the eight companies flew out to survey the condition of the fields and the refineries in Iran.

Protracted negotiations followed and in August 1954 the Iranian government representatives and the international consortium members announced full agreement in principle. The companies making up the consortium were the Anglo-Iranian Oil Company (which name was changed in December 1954 to The British Petroleum Company), Gulf Oil Corporation, Socony-Vacuum Oil Company (now Socony-Mobil Oil Company), Standard Oil Company of California, Standard Oil Company of New Jersey, The Texas Company, The French Petrol Company, and Royal Dutch-Shell Company; and it was their belief that in the first three full years of operation under this agreement Iran would net some $460,000,000. The agreement was to run for twenty-five years, with provisions for three extensions of five years, and the extremely technical text ran to nearly 50 pages. Ratified by Iran in October, tankers began loading at Abadan at once.

The consortium set up a British holding company called Iranian Oil Participants Limited which holds all the shares of two companies established to operate in Iran. These companies, registered in Holland, are the Iranian Oil Exploration and Producing Company, and the Iranian Oil Refining Company. In addition, the consortium established the Iranian Oil Services Limited. In 1955 one-eighth of the interest of the five American members of the consortium was transferred to and divided among nine other American oil companies: this so-called Iricon Group was made up of American Independent Oil Company, Atlantic Richfield, Getty Oil Company, Signal Oil and Gas Company, Standard Oil of Ohio, and Continental Oil Company. These participating companies deliver crude and refinery products to the National Iranian Oil Company at prices which include a fee above their operating costs. Each company established a Trading Company which purchases crude from the NIOC and either exports the crude or puts a part of it through the refinery at Abadan.

The consortium agreement provided that net profits were to be divided 50/50 through payments by the companies of a 50 percent income tax to Iran, while up to 12½ percent of the total crude could be, if so requested by NIOC, accepted as part payment of the profits. This later clause was intended to facilitate the entrance of NIOC into international markets. Compensation to AIOC included ten annual payments of 25 million pounds sterling for its fixed assets in the concession area, and by $510 million from the consortium in return for AIOC yielding 60 percent of its former rights.

Iran's revenues from the consortium mounted steadily: in 1956, $153.4 million; in 1960, $285.2 million; in 1965, $513.2 million; in 1966, $580.7 million; in 1967, $711.5 million; in 1968, $810.5 million; in 1969, $924 million; in 1970, $1,048.1 million; in 1971, $1,768.8 million; and in 1972, $2,400 million. NIOC does not share in these revenues which all go to the Treasury. As noted elsewhere, up to 80 percent of these revenues are turned over to the Plan Organization.

The National Iranian Oil Company was established in 1951 within the framework of the Nationalization Law and was authorized to explore, exploit, store, refine, sell and transport hydrocarbons, and to develop industrial and commercial hydrocarbon activities through subsidiary companies. Its role was broadened by the Petroleum Act of July 31, 1957, which permits it to form its own companies in the hydrocarbons field, or to form joint ones in which it holds not less than 30 percent of the ownership, or to enter into operating agreements with other companies. The land areas of Iran and its continental shelf in the Persian Gulf were divided into districts of not more than 80,000 square kilometers each, and NIOC was enabled to declare any district or part thereof open to offers from firms engaged in international marketing.

In August 1957 NIOC and AGIP (Assienda Generale Italiana Petroli) created a mixed organization named SIRIP (Société Irano-Italienne des Petroles). It was hailed as the first 75/25 percent to be concluded anywhere in the world: NIOC receives 50 percent of profits as a 50 percent owner, and an additional 25 percent in the form of 50 percent income tax on the profits of AGIP. The agreement called for a specified sum to be spent by AGIP on exploration, with this sum recoverable from SIRIP when commercial exploitation began, with half this sum going to each partner. The area was 22,900 square kilometers: one block inland, one offshore in the Gulf of Oman, and one offshore opposite Bandar Daylam near the northwest corner of the Gulf. By the twelfth year the area was to be reduced to places containing commercial deposits. The agreement was for 25 years from the first sales, with the possibility of three five-year extensions.

Oil was struck off Bandar Daylam in 1960 and this Bahregan field was declared commercial in 1962. Nearby the Hendijan and the No Ruz fields were brought into production at later dates. Near the village of Bandar Daylam is a tank farm with a capacity of some 2 million barrels. A marine terminal which accommodates 50,000 ton tankers has been augmented by an off-shore single buoy mooring facility capable of loading tankers up to 250,000 tons.

In June 1958 NIOC concluded the first joint structure

agreement when it formed the Iran-Pan American Company with the Pan American Petroleum Company. This company later assigned its interest to Pan American International Oil Company (now Amoco Iran Oil Company), a subsidiary of Standard Oil of Indiana. The Iran-Pan American Company (IPAC) was the operating agent. The joint structure was a non-profit company owned equally by the partners. The agreement had new features. Taxes were payable by the partners, not by the company. An annual rent per square kilometer at progressively rising rates was to be paid by Pan American. Pan American paid a cash bonus of $25 million which would be amortized in the years after commercial production began. Exploration costs of $82 million during the first twelve years were to be borne by Pan American, with exploitation costs shared by both parties. Each partner is entitled to half the crude available for export, and may purchase any part of the other half which is not taken by its owner. As in the SIRIP agreement, there was no provision for royalties. The agreement covered an offshore area of 16,000 square miles, covering a block which extended from the Iranian shore to the median line of the Gulf and which surrounded the island of Kharg.

In 1965 NIOC entered into agreements quite similar to that of the joint structure agreement with Pan American, covering areas of the continental shelf and forming companies whose names are abbreviated as DOPCO, IROPCO, IMINOCO, LAPCO, FPC, and PEGUPCO. Cash bonuses, often described as cash deposits, totalling $150 million, were paid by the foreign partners.

Dashtistan Offshore Petroleum Company (DOPCO) was formed with the participation of Royal Dutch Shell for an area of 6,075 square kilometers. There appear to be no later reports of its activity.

The Iranian Marine International Oil Company (IMINOCO) is owned 50 percent by NIOC and 16.67 percent by each of AGIP of Italy, Phillips Petroleum Company, and the Oil and Natural Gas Commission of India. The offshore areas comprise four blocks totalling 7,960 square kilometers. In July 1969 the Rustam field, some 65 miles southwest of Lavan island, came into production and an 18-inch submarine pipeline was laid to the island. The Rakhsh field, 17 miles to the north of Rustam, came into production in 1971. Near the end of 1972 these fields were producing some 80,000 barrels a day. At the Rustam field were two clusters of platforms, housing flow stations, a pumping station, and living quarters in 190 feet of water. Storage and loading facilities are on Lavan island.

The Lavan Petroleum Company (LAPCO) is owned 50 percent by NIOC and 12.5 percent each by Atlantic-Richfield, Murphy Oil Corporation, Union Oil Company of California, and Sun Oil Company. Atlantic-Richfield is the coordinator of the group. The offshore area covers 8,000 square miles. The Sassan field, 88 miles south of Lavan island, was discovered in 1965 and brought into commercial production in 1968. A 22-inch submarine pipeline connects the field and the island. Near the end of 1972 production was almost 200,000 barrels a day. Another field, some 52 miles south of the island, is under evaluation. The Sassan field is marked by four steel platforms in some 80 feet of water. Storage and loading facilities are on Lavan island.

The Farsi Petroleum Company (FPC) included NIOC and three French companies known collectively as SOFIRAN. These companies did not include the Compagnie Française des Petroles (CFP). The offshore area was 5,800 square kilometers.

The Iranian Offshore Petroleum Company (IROPCO) included NIOC and seven American companies: Tidewater, Superior, Skelly, Sunray DX, Kerr-McGee, Cities Service, and Atlantic. The offshore area was 2,250 square kilometers. There is no record available of the later activity of this company.

The Persian Gulf Petroleum Company (PEGUPCO) included NIOC and six German companies and covered an offshore area of 5,150 square kilometers. There appears to be no record of recent activity.

In December 1966 NIOC concluded a contract with ERAP (Entreprise de Recherches et d'Activités Petrolières—a state-owned French corporation) which was neither a mixed organization nor a joint structure. A non-profit company named SOFIRAN was established as a general contractor for NIOC. The area included 250,000 square kilometers onshore in the interior of Iran and 21,500 square kilometers offshore, approximately between Lavan island and Qishm island and extending south of the median line of the Gulf. ERAP finances exploration and development operations. The exploited crude is owned 100 percent by NIOC. ERAP has the right to purchase between 35 percent to 45 percent of this crude, and also acts as the broker for sales of certain quantities of NIOC oil. ERAP pays no royalties nor income tax, but an amount equivalent to an income tax of 50 percent is wrapped up in the price to be paid for the crude. The contract area shrinks very rapidly after exploration begins. Half of the recoverable reserves are set aside as national reserves. The contract runs for twenty-five years from the date of exploitation. When oil is discovered in commercial quantities exploration and development costs are recoverable from NIOC. In June 1972 oil was struck southwest of Sirri island.

In March 1969 NIOC concluded its second contract agreement with a group of five European companies: ERAP with 32 percent of the shares, AGIP 28 percent, Hispanoil 20 percent, Petrofina 15 percent, and O.E.M.V. (Austria) 5 percent. The contracting company is called EGOCO. The area is for 27,260 square kilometers in the province of Fars inland from the Gulf and including its territorial waters up to the three mile limit: the area is that released by the Consortium in 1967. The agreement is for twenty-five years. Within eight years after exploration begins 95 percent of the area must be relinquished. Fifty percent of recoverable reserves become national reserves, and the company may purchase 45 percent of the production below 275,000 barrels a day, and 30 percent if above that figure. It is also obligated to purchase certain quantities of NIOC's oil.

A third contract agreement was concluded with Continental Oil Company in April 1969. The area is 12,860 square kilometers situated to the north of Bandar 'Abbas in a region relinquished by the Consortium in 1967. The terms parallel those of the earlier contracts, with the exception of a cash bonus of $10 million from the company.

Two areas of the Gulf relinquished by the Consortium were opened to international bidding in the summer of 1970. In July 1971 the Hormuz Oil Company (HOPECO) was established by NIOC and Mobil Oil on a 50/50 partnership basis. The agreement is for twenty years, with two possible extensions for five years each, and the area is 3,200 square kilometers in the Strait of Hormuz. Exploration costs are borne by Mobil Oil and are only 50 percent recoverable. A cash bonus and a production bonus is called for. Income tax is levied at current Iranian rates, now 55 percent, and, in addition, a payment of 12.5 to 16 percent of the posted price of the exported crude is payable to NIOC. The terms of this agreement are more favorable to Iran than the comparable ones reached in 1965. The other partnership concluded in 1971 was with the Amerada-Hess Company and resulted in the Bushire Oil Company (BUSHCO). The area is 3,715 square kilometers offshore and close to Bushire and the terms are very similar to those reached with the Hormuz Oil Company. In June 1972 oil was struck southwest of Bushire.

NIOC has sale agreements with Spain, Romania, Yugoslavia, Czechoslovakia, Hungary, Poland, Argentina, and Japan. In 1972 NIOC sold 83.7 million barrels of crude oil for some $165 million.

In November 1969 Iran International Oil Company applied to the United States government for an allocation and license to import 250,000 barrels a day of Iranian crude. This company is a Delaware corporation and a wholly-owned subsidiary of NIOC organized to engage in the importation, refining and marketing of crude oil and petroleum products in the United States.

In early 1972 the British Ministry of Trade and Industry awarded a concession to NIOC and British Petroleum for the exploration and production of oil and gas in two regions of the North Sea. NIOC and BP have a partnership in the concession on a 50/50 basis, and to implement the project NIOC has established a British subsidiary, Iran Oil Company Ltd.

In addition to owning and operating refineries in the interior of Iran, NIOC has engaged in joint ventures elsewhere. In 1969 the $52 million Madras refinery came into operation in India as a joint venture of NIOC, the Government of India, and Amoco India Inc., a subsidiary of the American International Oil Company. India owns 74 percent and the other participants each 13 percent. The crude is supplied by NIOC from the Darius field in the Gulf. The annual capacity of the refinery is 2.5 million tons. In May 1971 the refinery at Sassolberg in South Africa came into operation, completed at a cost of $110 million. Its capacity is 50,000 barrels a day. NIOC has a 17.5 percent ownership and has contracted to provide 70 percent of its crude for 15 years from the Sassan field in the Gulf.

In the opening months of 1972 NIOC and members of the Niarchos group formed a company called the Irano-Greek Oil Refining and Distribution Company. The company holds two-thirds of the shares of the Aspropyrgos refinery, some 30 miles northeast of Athens, with the other third belonging to the Greek government. The refinery has an annual capacity of 4.5 million tons, and NIOC is to supply it with 32 million tons of crude during the next twelve years.

In December 1972 an agreement was concluded between NIOC and the Minister of Economy of Belgium for the establishment of a joint company, owned 50/50 by NIOC and Belgium, called the Irano-Belgian Refining and Marketing Company. Plans call for the construction of a refinery in Liége with an annual capacity of 5 million tons a year. NIOC will supply its total crude oil requirements.

The original Abadan refinery, on the island of that name and bordering the Shatt al-'Arab, was erected in 1911 with an annual capacity of 120,000 tons. Its capacity was expanded many times over the years. In 1972 it processed an average of 422,000 barrels a day in the form of aviation gasoline, motor gasoline, jet fuel, kerosene, fuel oil, and a number of other products. With the depth of water at Abadan only sufficient to receive smaller tankers the port of Bandar Mah Shahr on the Khor Musa, a deep tidal inlet, at a point about 45 miles from the open waters of the Persian Gulf and about 65 miles north of Abadan, was developed to handle tankers of up to 50,000 tons which could load both crude oil and the refined products piped from the Abadan refinery. With the appearance of larger and larger tankers, in 1958 the *Cham*, or Bend, project was undertaken. While Bandar Mah Shahr continued to be the loading point for refined products, the island of Kharg was selected as the crude oil port.

Kharg island is located on the deepest water line of the Gulf, 25 miles from the mainland, and has an area of 12 square miles. In 1960 the Kharg Island Maritime Terminal came into operation. More recent undertakings have made it the largest crude oil loading terminal in the world. In 1972 exports were at a level of about 4 million barrels a day. Five submarine pipelines run from major fields in Iran: so-called Iranian heavy flows from the Gach Saran field and Iranian light from the Agha Jari field. It is pumped to a tank farm with a storage capacity of 14 million tons. From the tank farm the oil flows by gravity to a jetty off the east coast of the island with ten berths for tankers up to 250,000 tons in 65 feet of water. In November 1972 an additional loading facility known as Kharg IV Project, or as *Azarpad*, was completed at a cost of some $37 million. This is a sea island about one mile off the west coast of Kharg in 105 feet of water. From the tank farm oil flows to the sea island in two 56-inch submarine pipes: its two berths can take tankers of up to 500,000 tons. In 1973 NIOC announced plans to construct a very large refinery on Kharg island.

The National Iranian Tanker Company is wholly owned by NIOC. There are several tankers in this fleet and others of 250,000 tons on order. The agreements with those companies operating on the continental shelf call for preferential consideration to the use of this fleet.

Within Iran the steadily growing demand for refined products is met by several refineries which receive crude by pipelines from the southern oil fields. Tehran has one with an annual capacity of 5 million tons, and a second of the same capacity is under construction. Shiraz has one of 2 million tons and Kermanshah has two refineries which obtain their crude from the Naft-i-Shah fields.

NIOC and the consortium companies signed a new agreement on May 24, 1973: it was effective as of March 21, 1973. This agreement replaced the existing one which was to run until 1979 with provision for three 5-year extensions. Its major provisions are as follows. The Iranian Oil Exploration and Producing Company was dissolved and replaced by Iranian Oil Services, registered in Iran and owned by the member companies of the consortium. The Iranian Oil Refining Company turned over the Abadan refinery to NIOC. The consortium is to raise its export capacity to 7.6 million barrels a day by October 1, 1976, and is guaranteed oil for twenty years at prices comparable to those in effect in the world's markets. The consortium will sell crude for consumption within Iran at its cost of production. In addition, it will supply NIOC with the following volumes of crude for export: 200,000 barrels a day in 1973, 300,000 in 1974, rising by 150,000 barrels annually to 1978, 1.1 million barrels in 1979, 1.3 million in 1980, and 1.5 million barrels a day in 1981.

The NIOC has stated that under this agreement Iran will supply 29 billion barrels of crude to the companies of the consortium during the next twenty years. Production, which in 1973 was 5.8 million barrels a day, will rise to 7.6 million barrels a day by October 1976 and then be stabilized at that level until 1984. After that date it will be reduced gradually to reach a level of 1.5 million barrels a day in 1993. The reason for this reduction is that known reserves, calculated at a maximum of 50 billion barrels, will approach exhaustion in twenty years, and after that time production will be largely for internal consumption. However, it would appear that the figure of 50 billion barrels relates only to the areas exploited by the consortium, and not for the areas of the continental shelf or other areas on the mainland.

In 1973 Iran was the second largest producer of crude oil in the Middle East: behind Saudi Arabia with 7.4 million barrels a day, and ahead of Kuwait with 2.7 million barrels, Iraq with 2 million, and Abu Dhabi with 1.3 million. About 50 percent of this crude goes to Japan, 25 percent to Western Europe, 12 percent to Asia, 10 percent to Africa, and under 3 percent to the United States.

The National Gas Company, a subsidiary of NIOC, is engaged in a variety of operations. By far the major one is the

Iranian Gas Trunkline. This project resulted from the agreement of 1966 between Iran and the USSR according to which Soviet credits advanced for the development of heavy industry in Iran would be paid for with natural gas. The line heads north from the southern fields, crosses the Zagros mountains at a height of 8,800 feet and then runs on to its terminus at Astara on the Caspian Sea within the USSR. The total length of the line is 700 miles, with pipe diameters from 40 to 42 inches: its cost has been estimated at $700 million. The gas supply contract calls for the delivery of some 90,000 million cubic meters in the period 1970-79 at a price equivalent to $6.60 per 1,000 cubic meters. In 1972 more than 8,000 million cubic meters were sent through the line. In 1973 discussions were underway for the construction of a second trunkline of like capacity. It was stated that the Federal German Republic would finance the construction of the line and in return receive a part of its capacity from connecting lines running through the USSR and on to Germany. From the existing trunkline, branches run to Ahwaz, Isfahan, Kashan, and Tehran, and supply gas for domestic consumption.

In addition to the huge gas reserves of the southern oil fields and the island of Qishm in the Gulf, a major field has been found to the south of Bushire. Another, at Sarakhs in northwestern Iran, supplies gas by pipeline to Mashhad, a distance of 75 miles.

The National Gas Company is actively promoting the foreign sales of LNG (liquefied natural gas). In September it signed an agreement with the Fuji Oil Company and the Marubeni Iida Company, both of Japan. It provided for the sale of four to six million tons of LNG for a period of twenty years, at a price delivered in Tokyo of 75 U.S. cents per million btu. The gas will be made available from Qishm island and will be utilized by the Tokyo Electric Company.

In March 1972 NGC exchanged letters of intent with the French ERAP group and the Japanese company C. Itoh for 3 million tons of LNG for a period of twenty years. The gas will come from the Gach Saran field.

LNG is produced, in part, in the Marun area, to the north

[279]

of the Gulf and between Agha Jari and Ahwaz. Production units supply NGL 400 and NGL 500 plants from whence the gas is piped to Bandar Mahshahr.

On 19 October 1972 an agreement was signed between the NGC and the International System and Controls Corporation of Houston, the Nissho-Iwai Company of Japan and the Norwegian companies Fred Olsen and Company and Halfdan Ditlev-Simonsen and Company as a joint venture. To implement the project the Kangan Liquefied Natural Gas Company (KALINGAS) will be formed and registered in Iran with 50 percent of its shares held by the National Gas Company. The cost of the project is estimated at $700 million which will be provided by the foreign parties. A plant to liquefy the gas will be built at Kangan on the Gulf, about 100 miles to the south of Bushire: the gas will come from offshore fields. Construction activities are to begin in the first half of 1973, with the plant completed in 1976. In its first phase it will produce about 1,300 million cubic feet of LNG a day and in its second stage about 2,000 million cubic feet. The capital of about $1 billion needed to build between 13 and 19 refrigerated ships to carry 125,000 to 160,000 cubic meters will be provided by the Norwegian companies and this fleet will be charted to the KALINGAS company. Initially the United States and Japan are envisaged as the principal markets.

In 1966 the National Petrochemical Company was established as a subsidiary of NIOC. It represented one segment of a many-pronged attack on the waste caused by the flaring of huge quantities of natural gas at the oil fields. It took over the Shiraz Chemical Fertilizer Plant which had been completed in 1963 at Marvdasht, to the north of Shiraz. Employing natural gas piped from the Gach Saran field, it produces about 55,000 tons of urea and 40,000 tons of ammonium nitrate annually. An adjoining complex produces carbonate and bicarbonate of soda.

The NPC was authorized to enter into partnership with Iranian or foreign companies to produce petrochemicals and to distribute, sell, transport and export these products. In each joint venture the share of NPC must not be less than 50 percent.

The Shapur Chemical Company was established in 1966 as a 50/50 joint venture between the NPC and the Allied Chemical Corporation. In 1970 NPC increased its share to 66⅔ percent and on 1 January 1973 NIOC took over complete ownership with Allied Chemical's equity to be returned over a six-year period. The plant was constructed on mud flats at Bandar Shapur at a cost of $232 million and included facilities for producing and piping sour gas from the Masjid-i-Sulayman field, some 108 miles to the north, at a rate of 60 billion feet a year. The gas has a hydrogen sulfide content of about 25 percent by volume. The complex went on stream in 1970 with the following anticipated annual capacity: sulfur 522,000 tons, ammonia 339,000 tons, sulfuric acid 429,000 tons, phosphoric acid 160,000 tons, urea 164,000 tons, triple superphosphate 71,000 tons, and di-ammonium phosphate 50,000 tons. The plant uses some 70 tons of imported phosphate rock each hour of its operations. The products are loaded at berths along a 1,800 foot long jetty. Plans for the development of this complex include a plant for the production of different kinds of petrochemicals. In October 1971 NIOC and the Mitsui Company of Japan for 50/50 participation in a plant to produce about 500,000 tons of olefins, including 300,000 tons of ethylene, 540,000 tons of aromatics, and 250,000 tons of caustic soda annually, as well as such derivatives as ethylene-dichloride, polyethylenes, ethylebanzen, and cumene. Investment is estimated at $358 million: the partners would each contribute $100 million with the balance in the form of loans and credits from financial sources in Japan. The natural gas would come from the southern fields, and the naphtha from the Abadan refinery. The complex is expected to be completed in 1976. The Shapur Chemical Company also plans downstream investment in conversion plants where the ammonia and other products exported would be converted into solid fertilizers.

The Abadan Petrochemical Company is a joint venture with NPC holding 74 percent of the shares and B.F. Goodrich Company 26 percent. The investment in the initial plant which went on stream in 1969 was $28 million. Annual production is P.V.C. (polyvinyl chloride), a raw material for plastics, 20,000 tons, and D.D.B. (dodecyl benzine), a basic material for detergents, 10,000 tons. Also, caustic soda 24,000 tons. A current addition to the plant will increase P.V.C. output to 38,000 tons. The plant uses gas from the Abadan refinery and locally produced salt. P.V.C. goes to an affiliated Polica plastics plant and to smaller plants throughout Iran.

In 1966 the Kharg Chemical Company was formed with 50/50 participation by NPC and Amoco International, a subsidiary of Standard Oil of Indiana. Construction was by the Japanese firm of Chiyoda at a cost of $44 million and the plant went on stream in November 1969. Sour gas comes from the Darius field and from a crude oil stabilization plant on the island. Its capacity is 220,000 tons annually of sulfur, and 2 million barrels annually of liquefied petroleum gases (LPG). The propane LPG is stored at 41°F. below zero and the butane LPG at 23°F. The LPG is sold to four Japanese companies on a ten-year agreement, beginning in 1969.

An agreement for the Madras Fertilizer Plant was concluded in May 1966 with 51 percent participation by the government of India and 49 percent by Amoco India Inc., a subsidiary of Amoco International Oil Company. The plant was completed in June 1971. In June 1972 NIOC exercised its rights in accordance with the original participation agreement to acquire 50 percent of the shares of Amoco India in the plant. The plant cost about $85 million and produces ammonia, urea, and NPK pellets.

In December 1972 a joint venture agreement was concluded between NIOC which was to have a 50 percent interest and the Mitsubishi Chemical Company and the Nissho Iwo Company, each with a 25 percent interest. It was for the investment of $40 million in a plant to be built adjacent to the Shapur Chemical Company. The plant, to be completed in 30 months, will produce about 40,000 tons a year of the plastic softening material di-octylphalate and such intermediate products as 28,000 tons of 2-ethylhexanol, 22,000 tons of phthalic anhydride and 32 million cubic meters of synthesis gas, the latter the raw material for 2-ethylhexanol.

Iran's share in the funding of the various petrochemical projects came from $400 million allocated in Iran's Fourth Development Plan, and it is anticipated that the later 1970's will see the total invested in such projects to be on the order of $1 billion.

The multiple activities of NIOC and of its subsidiaries require an ever increasing number of highly skilled technicians, administrators, and office workers. NIOC operates the Abadan Institute of Technology, founded many years ago by the Anglo-Iranian Oil Company, with 25 faculty members and 250 students. In addition to undergraduate courses, the Institute offers a graduate program in Engineering Sciences. In addition to the regular courses, which establish a fluency in general and commercial English, it provides a six-year course leading to a master's degree.

A Table of Organization of NIOC would indicate the ramifications of its supporting activities and services. Its Distribution Division, with 8,000 employees, services all Iran with petroleum products and natural gas. The Aviation Division services local airports and those at Kabul and Qandahar in Afghanistan. The Recruitment Department recruits and trains personnel and the Medical and Health Division looks after the employees. NIOC operates its own Microwave Communication System. It maintains an Advanced Administration Program at Tehran, and a Research Center, with a staff of over 100, at Ray, near Tehran. It also aids with the financing of an autonomous organization, the Iranian Petroleum Institute, which has scientific and technical committees, sponsors seminars and study groups, and issues publications. Finally, the principal offices of NIOC are in its own skyscraper at Tehran.

Cultivation and Soils

Over two-thirds of the land area of Iran is taken up by the highest mountain ranges and the great deserts and is therefore completely unsuitable for cultivation. According to official figures, 11.5 percent of the total area is under cultivation, but of this area, about 41 million acres, some 30 million acres are fallow during any year. Less than one-half of the average cultivated area is irrigated, or *abi*, land. The productive un-irrigated, or *daymi*, land is situated largely in the north-western and northeastern areas of the country.

Iran has little of the soil familiar to Europe and the United States, a loam containing living organisms resulting from the decomposition of natural vegetation. Instead, Iran's sub-stratum soil, a mixture of gravel, sand, clay and lime, has been exposed after centuries of deforestation and erosion. It is, however, quite fertile except when it is impregnated with alkalis.

Agricultural and Natural Products

Agricultural productivity and the growth of wild vegeta-tion is regulated by the scanty and concentrated rainfall of the country. The cereals, including wheat, barley, and rice, are the staple crops of the country, under cultivation on some 9,300,000 acres. Wheat is grown in every section of the coun-try except along the Caspian coast where its place is taken by extensive fields of rice which is also grown on the plateau in the few places where an abundant supply of water is at hand. Normal harvests are as follows: wheat, 4.4 million tons; rice, 1.4 million tons; and barley, 1.2 million tons. Wheat is the basic crop of the country and has been available in surplus until the scanty rainfalls of recent years have compelled Iran to import large quantities of it. Other field crops include corn, potatoes, millet, large peas, beans, and lentils. Alfalfa, native to the region, is a crop worthy of more extensive plant-ing since its long roots can penetrate to the sub-soil moisture. Cabbages, turnips, onions, eggplant, cucumbers, and melons are commonly grown, and also sugar beets, cotton, tobacco, and opium poppies.

Iran produces very fine fruits with the annual production about 450,000 tons. Most of them are not equal in size to our own highly developed species, but they have a more delicate flavor. Fruits grown include peaches, apricots, plums, cherries, pears, pomegranates, and apples. The peach and probably several other fruits originated in Iran and spread from there to other countries. Apricots are the quantity crop and are dried for home consumption and export. The apples are rather small in size, probably because the winter cold is not sufficiently prolonged to maintain the period of dormant life required by apple trees.

Citrus fruits, including oranges, tangerines, lemons, and limes, are grown along the warm and humid Caspian littoral and also in such southern regions as the town of Shiraz. The current annual citrus yield is 110,000 tons.

The date palm is cultivated in oasis villages inland along the entire length of the Persian Gulf and in large groves at the head of the Gulf. There are in Iran some 9,000,000 date palms whose present annual production is about 280,000 tons. In the south of the country the date tends to replace bread as the staple article of diet.

It has been estimated that some 400,000 olive trees are grown in a limited area on the northern slopes of the Alborz Mountains. Their crop of fruit and the oil it yields is one of real value, and attempts will be made to raise these trees in other parts of the country.

More than thirty varieties of grapes are grown on the plateau, the methods of cultivating them varying according to local habits. In one interesting method a circular pit is dug to a depth of several yards until moist earth is reached. The vine root is then planted at the bottom of the pit, and as it gradually grows toward the surface the pit is filled in with earth until the established vines are all above ground level. Grapes are a staple of diet during the summer months, are dried as raisins for export in amounts up to 60,000 tons a year, and are used in the making of wine. European types of wine are now produced at several towns, but the distinctive wine of Shiraz, resembling a Malaga in taste, is the most renowned.

Since the cereals are largely consumed in the areas where they are grown, a number of other crops form the principal source of farming cash income. These products are cotton, sugar beets, tobacco, tea, and silkworm cocoons.

Under the supervision of a government company which supplies seed and encourages production, the yield of raw cotton has increased steadily until over 500,000 tons are grown; exports have reached 85,000 tons.

Within recent years the consumption of sugar, said to be a reliable indicator of a fast growing economy, has increased at a fantastic rate. Domestic production has increased almost as rapidly so that the heavy imports of earlier years are no longer required. Nine state-owned and 21 private sugar refineries produce over 500,000 tons of beet sugar and well over 100,000 tons of cane sugar. Sugar cane plantations are expanding rapidly in the province of Khuzistan.

Tobacco is grown on at least 55,000 acres, principally in the areas of Gilan, Gorgan, Mazanderan, and Azerbaijan with seed tobacco produced in Mazanderan: annual production is about 15,000 tons. The purchase and processing of the crop is in the hands of a monopoly of the government which contracts with growers. The cigarette tobacco is of the Turkish type, and of the categories known to the trade as Trebizon, Tiklac, Basma, and Samsoun. The factory at Tehran turns out some 12 billion cigarettes annually, and efforts are being made to push the export of tobacco and cigarettes. Pipe tobacco production is at 3,600 tons and declining. Tobacco for use in water-pipes is grown in the Khonsar region and amounts are exported to Egypt and Syria.

Tea is the most popular beverage in Iran. A score of years ago tea plantations were developed in Gilan, and in more recent years Chinese experts were brought in to improve the quality of the product. The local production never exceeds 80,000 tons, primarily because the Persians prefer the taste of imported teas and will pay higher prices for them. Efforts to increase consumption of local tea, such as licensing local tea growers to import foreign tea in proportion to their own sales, have not proved too successful.

In summer the white fields of opium poppies have long been a familiar sign in Iran; the cultivation demanded relatively little attention and the cash return was high. It was less than a hundred years ago that the cultivation of the poppy was fostered in Iran in order to meet the demands of the European market, but the Persians themselves became victims of their own crops and, according to a 1955 estimate, there were 2,800,000 addicts in Iran. The preparation and sale of opium was a monopoly of the government and a very profitable one: in years when production ranged between 700 and 1,200 tons, an average amount of 80 tons was exported. In 1946 a cabinet decree was issued forbidding the culture of the poppy and the sale of opium, but in the years which followed the government continued to process and sell the crops. Strong sentiment against production grew, and in 1956 a bill was passed which prohibited the cultivation, production, and sale of opium, and provided stiff penalties for violations of the law. Health officials entered on a serious campaign to close down places where opium was smoked, including an estimated 1,200 at Tehran, and to cure addicts. More recently, as noted earlier, the production of opium has been renewed under strict government controls.

Mulberry trees grow all over Iran and for centuries many villages raised the silkworms which fed on the mulberry leaves and spun and wove the silk on hand looms, but it was not until 1885 that cocoons of higher quality were introduced from abroad. In 1937 the production of silk became a monopoly of the government, and centers for preparation of cocoons and for weaving were established along the coast of the Caspian Sea. Annual production ranges between 500 and 1,000 tons, and the industry does not reflect a stable situation capable of attracting additional producers.

Not much more than nine percent of the country is covered with timber, possibly 45,000,000 acres. The forests were nationalized in 1963. The dense forests which make up a full one-third of the wooded areas are found along the northern slopes of the Alborz Range, beginning a short distance back from the Caspian shoreline and extending upward to a maximum altitude of 7,000 feet above sea level. Much of this area is virgin forest, of oak, ash, elm, beech, ironwood, box, cypress, maple, and honey locust, and since it represents important potential national wealth all wood cutting is controlled by governmental regulations. Wood in the form of telegraph poles, timbers, and firewood is transported to the plateau by train and truck. A government factory near Babol on the Caspian coast cuts and creosotes ties for the Iranian railway system.

On the plateau proper the wooded areas are usually confined to the higher mountains and consist of sparse growths of scrub oak, whose bark is used in the tanning of hides, and of wild fruit and nut trees. The extreme deficiency of timber on the plateau is met in part by the cultivation of watered groves of poplar trees whose trunks are used in house construction. Willows and enormous plane trees grow in the villages and line the irrigation channels. Groves of fine black walnut trees are found throughout the country, and both almond and pistachio trees are much cultivated, the nuts from these trees constituting a good cash crop for both the nomad and the farmer. In fact, the income received from the harvest of a single giant walnut tree will support an entire family throughout the year. Some 40,000 tons of almonds and 9,000 tons of pistachios are harvested each year.

Deforestation of the plateau has been going on for centuries and still continues. The nomads cut the scrub oaks and burn the wood for charcoal, flocks of sheep and goats devour the new shoots, and everywhere the villagers carry on a constant search for fuel. Through the efforts of the Forest Service the heavy inroads of the charcoal industry have been slowed down, while nursery trees suitable for dry areas are now available at several centers for village planting.

Along with the nationalization of the forests came an extensive program of forestation. Travelling along many of the roads of Iran one can see groves of young pine trees spreading across large expanses of hillside. Even in such places as arid Bandar 'Abbas, if the saplings are watered for two or three years they grow without further attention.

Certain wild plants and shrubs and their saps or resins are carefully collected and form an important item of export. Some of them are gum tragacanth, gum arabic, gum asafetida, and galbanum, colocynth, and licorice. Some 2,500 tons of the medicinal gums are exported annually. Other plants of value include mastic, rue, absinthe, cumin, and sumac, and a score of herbs are widely grown to flavor rice and sauces served with rice. Coloring matters used in rug weaving come from indigo, saffron, and gall nuts. Henna for coloring hands and feet, hair and beard—used much less now than formerly—comes from the tree of that name. Jute is indigenous to the Caspian littoral where a maximum annual production of 6,000 tons of fiber was once reached. Oils are extracted by pressure from seed-bearing plants such as cotton, linseed, sesame, castor, and poppy.

Flowers are so highly prized by the Iranians that even the smallest courtyard is bright with flower beds and potted plants, and in early spring both valley floors and mountain slopes are carpeted with wild flowers. Some of those common to Iran are the tulip, grape-hyacinth, gladiola, gentian, bell-flower, poppy, buttercup, crocus, pink, iris, geranium, and dwarf hollyhock, and many wild flowers which are less familiar to other countries. Of them all, first place in popular affection is held by the rose, praised in charming verses by poets of nearly a thousand years ago. In fact, the common word for rose, *gol*, is also the generic term for flower. Many varieties are cultivated or grow wild along trails and watercourses. Noteworthy is the double pink, *rosa centifolia*, and the double yellow, *rosa hemispherica*, believed to have originated in Iran. Rosewater, *golab*, has been prepared in the country for centuries and at Kashan extensive fields of the double roses are still grown for that purpose. It is also of some interest to note that nearly 300 tons of dried roses are exported each year to countries such as Oman, Kuwait, Pakistan, and Iraq.

Water Supply and Irrigation

The fact that Iran's rainfall is very limited and that the fullest possible use must be made of the existing water supply has already been stressed, and it has been mentioned that although a good deal of dry farming is carried on in the northwest and northeast of the country agricultural productivity is dependent upon various methods of irrigation.

A system very effective for hillside farming is that in which dug channels angling off from rivers and streams carry water directly to the cultivated fields, the wider channels subdividing into narrow ditches. Parallel ditches at different levels along the slopes are connected vertically by miniature water-falls which can be left open or blocked by stones and brush. However, in this method a very large percentage of the original water supply is lost through seepage and evaporation in the channels and, of course, so much water is taken off near the source of the stream that little is left for areas further downstream.

The second and long the most important method of irrigation in Iran is by means of underground channels which in Persian are called *qanat* or *kariz*, a method used throughout the country, especially in areas where there are few flowing streams. The system seems to be almost uniquely Iranian, for it is seldom found outside of the country. Some authorities believe that the first qanats were dug in prehistoric periods. The tunnel of an ancient qanat, the only one known in Egypt, has been found in the oasis of Kharga, over a hundred miles west of the valley of the Nile. Since it is near a temple erected after the conquest of Egypt by the armies of Iran in the fifth century B.C., it seems highly probable that the invaders constructed there a system which was then common to Iran.

A qanat line is constructed to supply water for the needs of a farming village and for the irrigation of its cultivated fields. At the base of mountains, and at the point nearest in a straight line from the village, a master well or shaft is sunk deep into the ground until its bottom is below the summer water table. The water table is in a layer of porous rock over impervious clay which retains the water which seeps down through the ground after the warmth of spring has melted the snow on the high peaks. The master shaft is at least two hundred feet deep—near the village of Gunabad in eastern Iran are master shafts nearly 1,000 feet deep which are known to have been dug at least five hundred years ago. Often a series of horizontal galleries fan out from the base of the shaft so that water from a larger area of the water table can be drawn into the completed system. After the master shaft is finished, a trench is started on the slight slope above the village and aimed directly at the distant shaft, and when it reaches a certain depth it becomes a tunnel. As the tunnel is driven through the ground it is pierced by vertical shafts dug at intervals of fifty or more yards, which serve to bring air to the workers in the tunnel while excavated material is placed in baskets and drawn up the shafts by a rope and windlass. Finally, the tunnel reaches the master shaft, and water rushes through it to the village.

The excavation of qanat lines is in the hands of a limited number of specialists from families which have done this work for generations, the town of Yazd being the home of the most skilled qanat diggers. With simple equipment these men lay out the course of the tunnel and plan its slope so that it is just sufficient to keep the water in motion. Occasionally the tunnel is dug from the base of the master shaft toward the village. A qanat line may vary in length from a few hundred yards to a distance of fifteen or twenty miles, and it may take years to finish the digging of a single line. One line at Yazd is over forty miles long. The underground tunnel is about two and a half feet wide and four feet high, just room enough for one digger. In soft or sandy ground a collapse of the tunnel roof is guarded against by lining it with cylindrical lengths of baked tile. Once a qanat line is completed the work is not at an end, for it must be kept clear of silt and debris which has blown down the many vertical shafts. Sooner or later serious cave-ins take place. At first new sections of tunnel may be dug around such points but eventually the entire line becomes blocked and must be abandoned. Thus, when a village is seen from the air the paths of four or five qanat lines may be followed by the heaps of earth around the vertical shafts, but of this number probably not more than one is in active service.

A qanat line which taps a generous water table will furnish a flow of about four cubic feet a second, or an amount adequate for the periodic irrigation of about 200 acres. Water rents vary throughout the country but an abundant qanat represents an annual income of several thousand dollars. The qanat line emerges on the surface on the slight slope above the village, and at first is usually a single channel which supplies the power to turn the grindstones of the local mill. Beyond the mill the channel subdivides, and its branches flow first through the orchards and the grounds of those houses which stand in their own gardens, and then along the village lanes. Beyond the last houses they spread out to cover a fan-shaped area of cultivated fields.

The channels or runnels through the fields pass at right angles to rectangular plots of ground, each enclosed by a dyke of earth a foot in height. Each plot is irrigated by turning the water from the runnel into the plot until it is covered with an inch or two of water. During cold weather the fields of winter wheat receive three or four irrigations; in summer each plot is irrigated at intervals ranging from eight days to two weeks. Normally the problem of water rights is very complicated, for the output of a qanat may be the property of a single person or may be jointly owned by several people, each of whom possesses a different percentage of the total. One or more men are charged with the distribution and diversion of the water supply. The unit of time for measuring the flow to each plot is based upon the time required for a small concave brass bowl, pierced at the base with a tiny hole, to sink in a larger bowl of water.

The third means of irrigation is through water drawn from wells. Comparatively restricted use is now made of this means except in the southwestern corner of the country, where animals are employed to turn water wheels or to draw up leather buckets of water, both methods common to the countries of the Middle East and North Africa. It was not until the 1940's that attention turned to drilling wells of various depths which raise water for irrigating fields by means of gasoline powered pumps. Currently there are at least 35,000 such wells in operation with hundreds of them concentrated in the areas around Qazvin and Isfahan. There are many stations for servicing and repairing the pumps, and a factory turns out diesel motor pumps.

The fourth method of irrigation is by means of networks of canals led off from dams and barrages thrown across the rivers. The construction of such dams and barrages may date from Achaemenid times, although no surviving ruins can definitely be assigned to this historical period. It is known, however, that dams and their associated irrigation channels were in operation in southwestern Iran by the first century A.D. Remains of many dams and barrages built during the

Sasanian period, between the third and the seventh centuries A.D., have been found over a large area of south and western Iran, a few still largely intact and serving for current irrigation schemes. A most impressive series of dams was erected along the Karun and the Diz rivers, and today the air traveler flying over Khuzistan can easily trace the lines of the broad channels which led off from the reservoirs. After the Muslim occupation of Iran and down to the thirteenth century these dams were well maintained and the systems expanded, with the result that many thousands of acres were then under intensive cultivation which are now completely barren. Coming down to more recent historical times, a few large dams were constructed in the seventeenth century and two of these, one near Sava and one near Kashan, remain in fairly good condition.

Within recent years many major dams have been constructed, and they fulfill the additional functions of generating electric power and supplying water to a number of cities.

The first of these dams to be constructed was the Amir Kabir. Completed in 1961, the 590 foot high concrete dam spans the Karaj river which flows from the Alborz range; it supplies water and power to Tehran, as well as water for irrigation. In 1962 the Shahbanu Farah Pahlavi dam was erected on the Sefid river: it supplies water to vast areas of Gilan province. The Muhammad Reza Shah Pahlavi dam was an undertaking of great magnitude. Completed in 1962, its 665 foot high concrete arch has created a vast lake, while its generators produce 500,000 kilowatts of electric power. In 1963 the Shahnaz Pahlavi dam was completed on the Abshineh river near Hamadan, and supplies water to that city. The Farahnaz Pahlavi dam on the Jaji river to the northeast of Tehran came into service in 1967; it supplies additional water for Tehran.

The Zayandeh river which flows past Isfahan rises in the Kuhrang range. A narrow ridge divides its drainage basin from that of the Karun river and its tributaries which flow to the Persian Gulf. Joining the Kuhrang river, just across this divide, with the Zayandeh river was first thought of by Shah 'Abbas early in the seventeenth century and thousands of workmen labored to cut a V-shaped trough through the ridge but abandoned the effort when it was only about a third completed. A few years ago this project was revived; a tunnel was driven through the divide and the waters of the two rivers merged behind the Shah 'Abbas Kabir dam, completed in 1970. In the same year the Aras dam, on the river of the same name which forms the frontier between Iran and the USSR in the northwest, was completed as a joint project of these neighbors.

In 1971 the Cyrus Kabir dam was constructed on the Zar-rineh river to the south of Mianduab in Azerbaijan. In the same year the Shapur the First dam was completed on the Mehabad river in the same province. Also, in 1971 the Darius Kabir dam was completed on the Khor river some 60 miles to the north of Shiraz; it supplies water to that town. The Reza Shah Kabir dam, on the Karun river, is the largest in the country, generating one million kilowatts of electric power. These dams supply water to irrigate over 800,000 acres of farm land, with the water sold at an average price of six cents a cubic meter.

The Water Resources Nationalization Law was enacted in 1968. It states that all waters flowing into rivers, natural streams, valleys, and other natural water surfaces, and all waters of natural lakes, springs, mineral waters, and subterranean water deposits are considered natural wealth and belong to the people. The Ministry of Water and Power has the duty of protecting and utilizing this national wealth, and of establishing and administering installations for the development of water resources. The building of improvements on the shores of rivers, natural lakes, reservoirs, and seas must have the prior approval of the ministry.

Fisheries

The waters of the Caspian Sea have long been a source of food and of income to Iran. In 1927 a joint Irano-Soviet fisheries company was given a monopoly on the foreign sales of fish and caviar from the waters off the Persian coasts of the Caspian. Although the company was owned in equal shares, the Soviets managed to dominate the management of the fisheries. In January 1952 the concession expired and Iran failed to respond affirmatively to a Soviet request that it be extended. Internal sales and exports are now in the hands of an Iranian company, *Mahi Iran.*

The Caspian Sea contains sturgeon, salmon, perch, mullet, whitefish, carp, and many other kinds of fish. While the USSR may take up to 400,000 tons a year in its waters, the Persians catch only about 7,000 tons. With a new trawling fleet in operation, Iran's catch should rise to 30,000 tons. A very large fish breeding farm established near Rasht specializes in producing sturgeon fingerlings. The sturgeon spawn in the clear cold waters of the streams which flow from the Alborz range into the Caspian, and are the source of caviar. The Beluga sturgeon generally weighs from 180 to 250 pounds and yields between 40 and 50 pounds of caviar. Two other varieties of sturgeon are considerably smaller and yield from 10 to 18 pounds of caviar. The production of caviar is over 200 tons a year. About 30 tons are consumed in Iran, 100 tons are sold to the USSR and are marketed abroad as Russian caviar, 60 tons go to the United States, and the remainder to Europe: the foreign sales amount to $4 million each year.

The fisheries resources of the Persian Gulf demand greater exploitation. Small vessels supply a fish canning plant at Bandar 'Abbas which packs sardines and tuna. There are many other varieties, including Spanish mackerel, Indian salmon, sea bass, rock cod, and plaice. Iran licenses foreign companies to trawl for shrimp in its waters and in 1971 put a new fleet of its own in operation, comprising several trawlers and a mother ship equipped to freeze the catch.

Farming Methods

Until very recent years plowing, sowing, and harvesting were carried out by the methods in use throughout the Middle East for thousands of years. The area of land which can be cultivated by a pair of oxen is called a *juft* or "pair," and requires about 2,000 pounds of wheat or barley seed.

The ground is broken by a team of oxen pulling a plow fashioned from a forked tree limb, with the plowshare encased in iron. Donkeys, or even camels, may take the place of the oxen, and the lightweight, modern steel plough is used with increasing frequency. Sowing and weeding are done by hand. The ripened wheat and barley is cut with sickles and tied into sheaves and—just as is recorded in the Bible—the poorer people have the privilege of gleaning the fallen stalks.

The sheaves are carried to the threshing floor, a hard-packed surface of clay located on the edge of each village. There the grain is heaped in piles and threshed by driving oxen, pulling a heavy frame lined with projecting teeth, in a slow circle over and over each pile. Winnowing is accomplished by pitchforking the grain into a strong wind. As the final step the grain is passed through a coarse sieve and the rounded piles of grain impressed with an owner's mark, a decorative motive cut into a block of wood. Flour for local consumption is ground in the village mill while the rest of the grain is sacked and transported by donkey or camel to the largest towns. The government also uses a fleet of trucks to collect and haul the grain to the large silos situated adjacent to the largest towns.

Fertilizers were sparingly available. In many areas the farmers dug at the mounds of ancient settlements and scattered the material over their fields. Animal manure was used primarily for fuel, but quantities were mixed with straw and spread on the fields. Animal manure and human waste was also used on plots of melons and vegetables. On the extensive plains around Isfahan a feature of the landscape is the high circular towers with interiors divided into hundreds of compartments for housing pigeons whose droppings are collected and used on the fields. However, the principal method of restoring fertility to the soil is allowing it to lie fallow. Most villages have far more land than can be irrigated with the existing water supply, and so a field which is cultivated one season is permitted to lie fallow for the next year or two. Crop rotation is also practiced.

Chemical fertilizers had to be imported and were beyond the reach of the average cultivator. Now, however, with the construction of several petrochemical plants, the supply is abundant and far less costly. Figures of doubtful reliability suggest that 100,000 tons of these fertilizers were used in 1969, and 285,000 tons in 1972.

Agricultural machinery, such as tractors, disc plows, and combines, came late to Iran. The first effort to farm a large area with machinery was made near Bushire after World War I, and ended in failure. The basic system of dividing the fields into small units for convenient flooding, together with the fact that they were often on sloping ground, made the use of tractors impractical in many regions. Extensive, flat, and unobstructed areas were needed for effective use of machinery, and the Gorgan plain to the east of the Caspian Sea met these requirements. In addition, the area was easily accessible by rail and road to Tehran. Large-scale farms were established after 1948 by local and outside entrepreneurs. The Gorgan Dry Farming Company operated over 5,000 acres, the Muhammadi farm is about 7,000 acres, and the farm of the Iranian Army is some 10,000 acres, while there are many of over 150 acres. Much of this land was at first leased, and later purchased, from the Crown Lands and included large areas which had never been farmed, as in earlier periods they had been the grazing lands of the Turkomans. The rainfall was sufficient for dry farming, and at first wheat was the principal crop. Then it shifted to cotton which brought a higher cash return. Tractor drivers, mechanics, and farm laborers were brought in from other areas of the country, and paid daily wages. It was soon discovered that irrigation would more than double the per-acre yield of cotton, and the owners invested in the drilling of wells and the digging of qanats.

In 1955 the Development and Resources Corporation of New York, headed by David E. Lilienthal, Sr., contracted with the Plan Organization to plan and execute a unified program for the agricultural and industrial development of the province of Khuzistan, including a system of dams and canals to provide water for some 175,000 acres. The Muhammad Reza Shah Pahlavi dam is a key unit in this plan. Canals and ditches were dug across a vast plain, burning hot in the summer, and an initial venture was the 33,000-acre mechanized sugar cane plantation at Haft Tepeh. It became apparent that agricultural production by local farmers was not increasing at the anticipated rate, and, as a result, in 1968 a law was enacted to authorize the establishment of companies for the purpose of maximum utilization of water resources and land irrigable from dams and irrigation installations on land downstream of the dams. Private foreign companies, financed by banks, corporations, and manufacturers of farming machinery, appeared and were ready to invest the $400 per acre needed to prepare the land for mechanized farming. These firms include the Iran-California Company, the H&N Agro-Industry of Iran and America, and the Iran-Shellcut Company, the latter with Dutch and British interests. And some 75 percent of the area has been allocated to these and other so-called "agribusinesses."

The number of tractors employed in agriculture has risen from 17,500 in 1968 to 24,000 in 1972. A tractor plant constructed at Tabriz by Romania assembles 5,000 vehicles a year, a figure that is to rise to 10,000. A second tractor plant at Arak is a joint enterprise between Iranian investors and the John Deere Company. The machine tool plant at Arak turns out agricultural machinery. The Ministry of Agriculture maintains tractor service and repair centers and finances the sales of tractors to farmers.

Animals and Birds

According to official estimates, the domestic animal population is as follows:

Sheep	29,000,000	Horses	600,000
Goats	12,500,000	Water buffalo	250,000
Cattle	6,000,000	Camels	400,000
Donkeys	2,100,000	Pigs	55,000

Over several years there has been an actual decline in the number of sheep and cattle, so marked in the case of sheep that large quantities of dressed and live sheep are imported. This situation has been blamed on successive years of scanty rainfall, and on the fact that agricultural companies, such as those of the Gorgan plain, have taken over large areas of former grazing land. The Ministry of Agriculture hopes that the solution lies in its establishment of several very large animal husbandry farms which breed livestock and grow fodder so that the animals are not dependent on grazing.

The breeding of fine Turkoman, Arab, and Persian horses, which are not used as draft animals, has been carried on largely by the tribes for many centuries. When Alexander the Great came to Iran he made a special trip to the mountains south of Kermanshah to see the famous Nisaean herds. Oxen draw the farm implements. The cattle are not systematically fattened for beef: meat is not a principal article of diet for much of the population, and when eaten it is usually mutton which is preferred over beef. Pigs are raised near Tehran and Hamadan for consumption by non-Muslims. In every farming community the donkey is the universal beast of burden.

The animal production of wool runs to about 35,000 tons, but exports which formerly amounted to 10,000 tons a year have fallen off sharply, and, indeed, large quantities are now imported. Lambskins, hides, and animal hair are important items of export income. Camel hair, in an annual production of 300 tons, is especially valuable.

The wild animal population of Iran defies all census-takers, and it is possible only to enumerate the species found there. The lion, the national animal of Iran, is now extinct. Tigers of great size are common along the Caspian. Brown bears live in the Alborz Mountains and smaller cinnamon bears in the Zagros Range. Panthers, jackals, wolves, and foxes are common. Porcupines, squirrels, hares (but no rabbits), rats, and mice abound. There is a variety of game: wild sheep and goats in the mountains, gazelle on the plains, wild asses in the salt deserts, and wild pigs in every swampy spot. Of the reptiles, turtles and lizards are numerous, but snakes are less common and only the horned viper of southeastern Iran is poisonous.

Many of the birds of Iran are common to the United States. An incomplete list includes the crow, raven, magpie, jay, oriole, finch, sparrow, lark, wag-tail, warbler, thrush, robin, woodpecker, kingfisher, and owl. Also the green-coated bee eater, turtledove, pigeon, partridge, quail, and pheasant.

Storks build their nests on the tips of the village shrines, and these welcome guests are known as *Haji Lak Lak* or "the pilgrim who cries lak lak." Small children call to the stork and he replies:

> Haji lak lak dar kojai?
> Dar bolandi
> Cheh mikhuri?
> Mikhuri nan qandi

> Haji lak lak from where?
> From afar
> What do you eat?
> I eat sugared bread

Ducks and geese breed in Sistan province. Small hens scrabble for a living in the dirt and debris of the villages, although there are many modern poultry farms near the urban centers.

Falcons are found inland from the Persian Gulf and are still bred and trained for hunting, principally by the nomads. The sport of falconry has long been popular in Iran and was a favorite of the ancient rulers; the birds were highly prized, some of them being valued at the equivalent of several thousand dollars.

Hunting in Iran features local techniques of proven effectiveness. Gazelle abound over the plateau and are a real challenge to the hunter on horseback. Sighting the horsemen, the gazelle takes off in one direction and will not swerve from that line, so that the riders follow an angle which may intercept the animal's course. Hunting by jeep, introduced by Allied troops during the war, has been the decay of the sport. Such "hunters" have timed the speed of gazelles as high as 50 miles an hour. The pursuit of mountain sheep has its own procedure. Along the lofty game trails holes are scooped out and the hunters lie concealed behind ramparts of earth and stones, while beaters attempt to drive the sheep into position.

Hunters of animals and birds must obtain licenses. Certain species are protected, such as the rarer kinds of gazelle and deer, and require special licenses. Hunting of tigers, cheetahs, and the famed wild ass of the salt deserts is prohibited. There are nine wildlife parks and some thirty protected regions which are not open to hunters.

Villagers and farmers employ primitive but efficient means of catching quail. A man will hide beneath a brightly colored patchwork quilt and, as the curious birds alight to examine the lure, thrusts out his hand through a hole and grabs the birds. In certain wide valleys the quail have customary stopping places in their flight across the valley. The villagers spring up from behind blinds and bring down the birds by swinging long bamboo canes; the Tehran game market is supplied by these hunters.

Persian cats deserve separate mention but the fanciers of this type may be surprised to hear that the Persians do not claim it as their own but assign it to Angora (in Turkey). Called the *gorbeh boraq*, or cat with bristling hair, it is not common in Iran except in the town of Yazd. There, it is said, quarters of the town breed their own lines, isolating each from contact with the others. Hairs from the tails of these cats form the delicate brushes used in miniature painting.

VIII. THE DEVELOPMENT OF FACILITIES

Industry

WHEN Reza Shah opened the first session of the VIIIth
Parliament in December 1930, he said: "We wish this
Parliament to be known in the history of the country as the
'economic parliament.' " From that time on every effort was
made to make Iran as self-sufficient as possible, and the government began the task by assuming the role of the "supreme
economic organizer." Industry within the country was to be
developed on a large scale but certain measures had to be
taken before the factories could be erected and put in operation. The world-wide depression had been acutely felt in
Iran, and it was necessary for the government to take over
strict control of foreign exchange transactions and to supervise the import trade so that only prime essentials would be
purchased with the scanty stock of foreign exchange. The
theory and operation of stock companies had to be explained
to the merchants, to the new industrialists, and to possible
participants. Further, stock companies formed to deal in articles of export and import had also to be prepared to distribute
the products of the factories then under construction.

A Ministry of National Economy was set up to regulate the
fields of agriculture, commerce, and industry. In 1931 a law
was passed requiring the registration of all stock companies,
and the compliance by these companies with certain regulations. In 1932 a new commercial code was put into force. The
government took the lead in the formation of new companies,
some of which were owned outright by the government, in
others the government had a controlling percentage, and in
still others ownership was divided between the stockholders
of the new company, the government, and one or more of the
companies already in existence. The operations and ramifications of some of these companies seem worthy of mention.

The Imperial Company was founded in 1931 with a capital
of 5,000,000 rials represented by 5,000 shares of stock of

which 2,000 shares belonged to Reza Shah, 2,000 to the National Bank, and the remaining 1,000 shares to two German firms manufacturing machine tools and electrical equipment. The Central Company, capitalized by the Ministry of Finance and the Agricultural Bank, was established to carry on commercial transactions abroad. It also had a controlling interest in the Company for Cotton, Wool, and Hides, which was to process and export these items.

The Cotton Goods Company, which had a paper capital of 20,000,000 rials, of which 52 percent was held by the National Bank and the rest by the Ministry of Finance and the company directors, was to have a monopoly on the import of cotton piece goods and their sale to private merchants. The merchants were to pay to the company 20 percent over and above the cost of the goods, and of this profit 15 percent was to go to the government as payment for the monopoly rights and 5 percent to the company as commission.

In August 1936 control over nascent industry was strengthened by the passage of a bill which stated that an individual or company which wished to establish a factory or industrial enterprise should make application to the Administration of Industry and Mines. Three categories of enterprises were specifically encouraged: those planned to supply vital local needs, such as sugar, woven goods, and matches; those planned to prepare and package important export items, such as cotton, rice, and fruit; and those associated with the trend toward urbanism, such as cement, glass, and railway ties. Remembering the earlier intervention in internal affairs resulting from loans obtained from Great Britain and Russia, foreign participation was not invited.

In 1936, firms in which the government owned blocks of shares or a controlling interest had a monopoly on the export or import of the following items: silkworm eggs, sugar and matches, opium, silk, cotton piece goods, jute, rice, playing cards, goat and sheep skins, saffron, asafetida, rugs, silk stockings, wool, alcoholic drinks, canned food, dressed skins, handbags, shoes, automobiles and trucks, tires and spare parts, and dried fruits. This meant that the government had direct control of 33 percent of all imports and 44 percent of all exports.

Profits to the companies were large: one company offered shares for sale with a guarantee of annual dividends of 12 percent; another spinning and weaving company made profits ranging from 40 to 50 percent. Favored government officials, merchants, and members of influential families profited materially from these companies but as time went by management was not willing to replace wornout machinery and to promote such efficiency that the local products could compete in quality and price with a growing flood of imports. Typical of the result was the announcement, in 1956, that the important Risbaf textile factory, founded at Isfahan in 1933, had gone into bankruptcy. The profits accruing to the government itself were used in part for the erection of factories and industrial plants. There was, of course, a temptation to set the prices of monopolized imports fairly high, and as a result the people of Iran paid for the industrial development of the country by a form of indirect taxation.

The system of monopolies and controls entered a new phase about 1937 with the conclusion of barter agreements between Iran and Germany, and between Iran and Soviet Russia. Russia sent piece goods, sugar, and other items in exchange for wool and rice.

With the advent of World War II the monopoly system played an even more vital role in the economic life of the country. Limited sources of supply and scarce shipping space made it necessary for the Iranian government to import, often through the medium of Allied agencies, none but the most essential items, and the institution of war-time rationing was also a factor which kept the government in business. Since 1942 a number of the monopolies have been brought to an end, although tea, sugar, and tobacco remain in that category.

Naturally enough Iran had lagged far behind the Western world in the development of the new machine age. At the end of World War I there was little industry within the country except for a few electric light and power plants and some match factories. Before Reza Shah's detailed program of industrialization was interrupted by World War II it had resulted in the erection of some thirty moderately large factories owned and operated by the government, and nearly two hundred other industrial plants. The major emphasis was placed upon textile weaving, with food materials and vegetable products in second place.

The general industrial scheme was a sound one and was executed by foreign specialists and by Iranian engineers who had been trained abroad. Plants were well located in reference to the new railway lines, to mineral deposits, and to areas suitable for the growing of the agricultural products to be processed. Many of the factories were designed by leading European engineering firms, and fine modern machinery was purchased from England, Germany, Sweden, and other countries.

Since the end of World War II industry has developed at an accelerated pace, aided first by funds supplied by agencies of the United States, and then financed by the budgets of the Plan Organization, by funds from ordinary government revenues, and by local and foreign investors. In recent years the emphasis has been three-fold: on heavy industry, on the production of durable consumer goods which had been imported, and on the stepped-up output of construction materials.

The most important segment of heavy industry, that which exploits natural resources, has been considered in the previous chapter. A machine tool plant at Tabriz, set up in collaboration with Czechoslovakia, produces centrifugal pumps, electric motors, diesel engines, and drilling equipment. The machine tool plant at Arak, established in cooperation with the USSR, manufactures boilers, cranes, and agricultural, industrial, and mining equipment. Tractor plants operate at Tabriz and Arak.

The rising output of durable consumer goods is clearly reflected by the automotive industry. Over 35,000 automobiles, 10,000 trucks, and 4,500 buses and station wagons are assembled annually by thirteen companies, most of which include investments by foreign manufacturers. The makes include Jeeps; Land Rovers; Mercedes-Benz, Mack, and Leyland trucks; Volvo; Citroen; and Peykan—a local name—cars. Currently progress is being made in changing over from assembling imported parts to the complete local manufacture of engines and major parts, notably in the Iran National Plant which produces the Peykan. These same companies are actively engaged in selling buses and trucks to nearby countries. Imported vehicles amount to only about ten percent of annual sales. Until recent years the majority of these imports came from the United States, but now Japan is the leading supplier of cars.

A plant at Qazvin manufactures 100,000 bicycles and 50,000 motorcycles a year. Three plants, each one sponsored by a foreign manufacturer, produce tires and tubes, with Iran well on the way to meeting the entire local demands in this field. A ball bearing plant supplies the automotive industry and other plants.

Numerous mills turn out over 11 million yards of woolen textiles and about half a million yards of cotton and artificial fiber textiles a year, as well as more than one million blankets. Some 75 million pairs of shoes are made annually, and find markets in other countries.

The Pars Paper Mill at Haft Tepeh uses as its raw material pulp from the nearby beet sugar refinery. Some 15,000 workers are employed in factories which manufacture air conditioners, electric and gas water heaters, space heaters, gas stoves, refrigerators, and electric fans. Without citing the rather impressive production figures, it should be pointed out that scarcely any of these items were produced locally a decade or so earlier. And also that they are now exported in considerable numbers.

Cement is the most essential item of the construction industry. Some twelve plants, located throughout the country, manufacture nearly 4 million tons a year, an amount which almost meets current consumption. Brick making is centered at Tehran: a group of kilns on the southern edge of the city, each marked by a soaring brick chimney, taller than any of the minarets which they so resemble, produce over 2 billion bricks a year. It is of interest to note that a number of years ago the type of brick used in this part of the world for many centuries—square in shape and less than two inches thick—was replaced by the European type. Sheet glass is manufactured, and a very flourishing industry produces the glazed tiles which have been so long a decorative feature of the mosques and other structures.

In those fields of production not within the major categories mentioned above, the pharmaceutical industry is very highly developed with some eighty plants turning out 2,500 different medical items. Most of the raw materials for these medicines are imported. Local chemical plants produce soaps and detergents, paints and dyes, and a very long list of other items for household and industrial uses. Food processing plants make vegetable oils for cooking, canned fruits and vegetables, candy, jams and jellies, biscuits, and dairy products.

About 6,000 large and small factories employ 300,000 workers, and 225,000 smaller industrial workshops employ 1.7 million people. The total investment in these enterprises is some $4.36 billion. Two-thirds of these enterprises are centralized at Tehran, but the current trend is to set up new factories in the provinces, as indicated above. In addition, industrial areas have been established in areas near provincial cities. The first two were near Ahwaz and Qazvin. Each such area includes several to a score or more factories, an administration building, housing, and necessary services.

The Institute of Standards and Industrial Research of Iran was established in 1953, and by 1973 had set standards for some 800 industrial and agricultural products and consumer goods made internally. Local criticism of the industry persists. It is charged that the overprotection of certain industries by the banning of competing exports results in higher prices and products of inferior quality. Local concerns may attempt to attract discriminating buyers by misrepresenting their products as foreign imports. Inadequate packaging may result in retailers receiving scratched or damaged air conditioners and other items.

Public participation in the ownership of industry was promised for a number of years, until in 1972 the government announced that shares in nearly all the state-owned industries would be offered to the public, as well as shares in a large number of privately-owned industries. The Industrial and Mining Development Bank of Iran, a heavy investor in development projects and industry, has some 6,000 shareholders. Then, in 1973 the National Iranian Oil Company announced that it would offer shares to the public.

Foreign Investment

After World War II, as part of a general reaction against foreign influences, regulations were put into force which appeared to penalize the operation of foreign firms in certain fields. New regulations concerning foreign banks appeared so restrictive that all such institutions except the Soviet-owned Russian Bank in Iran withdrew from the country. A cabinet decree of December 1952 listed rather stringent conditions under which foreign insurance companies could operate, including the withdrawal from the country of no more than 10 percent of annual revenues, and these concerns began to wind up their affairs.

Quite recently there has been a swing in the other direction as the need for foreign capital in certain fields has become apparent. Near the end of 1955 the Parliament passed a bill of seven articles entitled For Attracting and Protecting Foreign Capital. The bill provides for the withdrawal of the capital in the same currency in which it was brought into Iran, for taking out annual net profits in that currency, and for equitable compensation in case of nationalization. However, the maximum allowable net profit is to be determined by regulations, and in the case of nationalization the amount of indemnity is to be determined by a local commission.

Industrial enterprises may be granted a 50 percent exemption on income tax on their local profits, and they pay a 25 percent tax on their undivided profits. There is no legal requirement that a foreign firm may have less than total ownership of an industry or business in Iran. However, the Foreign Investment Promotion Center, operating under the Ministry of Economy, which reviews and passes on applications for investment in Iran may stipulate that in the case of joint stock companies there must be at least 51 percent local investment.

From 1956 until the end of 1970 foreign investments approached $100 million, with more than half this amount contributed by thirty-seven American concerns. Considerably lesser sums came from Great Britain and the German Federal Republic, and from a number of other countries. In 1957 Iran and the United States exchanged a note which provided for the latter country to guarantee private American capital investment in Iran; in cases of claims against Iran there would be direct negotiations between the two countries. In descending order of magnitude, the total investments were in petrochemicals, rubber products, pharmaceuticals, mining, metallurgical industries, and electrical appliances.

Several hundred firms engaged in industry, industrial sales and related businesses have offices in Iran: they include over 200 American firms, over 100 British firms, and nearly 100 West German firms. Also, with a lifting of earlier restrictions on banking, many foreign banks have established branches in Iran.

Rug Weaving

During the sixteenth century the art of rug weaving in Iran reached a height never equaled in any other country. Although the weaving of pile rugs and carpets was a national industry for centuries, the earliest preserved fragments date from the end of the fifteenth century. Similar rugs are pictured in Persian miniatures of the fourteenth and fifteenth centuries and in European oil paintings of this same period.

The hand looming of fine rugs continues in Iran on a large scale, and the preparation of the materials, the techniques of weaving, and the choice of patterns and colors follow the earlier traditions. The people of the country cherish these rugs as their most prized possessions. In addition, Persian rugs represent a most important item of export, and up to the present have been able to compete successfully with the machine-woven rugs of western countries. Fairly exact copies of fine Persian rugs are woven in Iran and other countries by power-driven looms, but they lack the vitality and luminousness of the original models, perhaps because the very irregularities of their pattern outlines and colors are the source of their beauty and brilliance.

Most of the rug weaving is done in so-called "factories." A factory may be a large house in which two or more looms have been installed, for there are comparatively few structures which have been specially built to house looms. There are said to be 120,000 of these vertical looms, with 400,000 regular and part time weavers. Most of the weavers are women. In 1965 a law was enacted which forbade the employment of children under twelve as weavers; prior to this time about 67,000 children below ten years of age had been spending long hours at the looms, as their small fingers were more nimble and supple than those of the older weavers.

Usually the rugs are woven on a base of cotton threads and the pile is made up of strands of wool or silk yarn which are knotted around the base threads and then cut off. Cotton is used in preference to wool for the foundation threads because the knots can be tied closer together on the thin cotton strands. Fairly coarse rugs may have some sixty knots to each square inch, while those of fine quality have over two hundred. The fabulous rugs surviving from the sixteenth century have nearly four hundred knots to the square inch, and even today this figure is sometimes excelled in the very costly silk rugs. The number of knots to the square inch has no relation to wearing qualities, but the finer workmanship gives an extra precision to the outlines of the pattern elements.

The quality of rugs intended for export is controlled, and weavers are encouraged to use only permanent dyes. Alizarin dyes are used to a certain extent but at most of the weaving centers the prepared yarn is boiled in vats containing such time-tested local coloring materials as madder, indigo, cochineal, and dyer's weed, and the skins or shells of the almond, walnut, pistachio, and pomegranate.

In the factories a preliminary drawing is made for each rug as well as careful full-scale pictures in color of sections of the pattern. In weaving a large rug three or four weavers sit in front of the loom while a foreman, following the prepared drawings, calls off the number and color of the knots to be tied by each weaver as the work goes forward.

The rugs are usually finished while still on the looms. As a section about two feet long is completed the foreman carefully clips the pile down to a uniform surface. Work progresses slowly: a rug of moderate size takes about a month to weave, and huge rugs, as large as 15 feet by 35 feet, which are sometimes woven, may take over two years to complete.

The rugs are usually known to the trade by the names of the towns or districts where they are woven, and many towns have their own patterns and combinations of colors by which their rugs may easily be distinguished. Trade names in common use include Tabriz; Heriz (near Tabriz); Qashqa'i and Afshar from the vicinity of Shiraz; Senneh, Bijar, and Sanandaj from the region of Kurdistan; Saraband from Hamadan; Mahal and Sarouk from the region of Arak; and the well known Kerman, Kashan, Qum, Isfahan, Na'in (near Isfahan), Mashhad, and Turkoman types.

A great many rugs are also woven by the women of the nomadic tribes or of the farming communities, and these are not usually exported. No preliminary drawing is made and the bold geometric designs develop as the weaving progresses. The women of the tribes weave on flat looms, stretched out on the ground, which may be quickly disassembled for transport as the tribes move from one region to another.

The Iran Rug Company was founded in 1936 as a governmental monopoly to supervise production and export sales. Currently its responsibilities are more restricted, although it buys extensively from weavers under contracts and maintains looms at such towns as Kerman, Arak, Kashan, Malayer, Hamadan, Mashhad, and Tabriz. This company attempts to preserve traditional patterns and to maintain traditional quality.

The output of rugs mounts steadily from year to year, and the increasing production is accompanied by certain problems. The supplies of factory woven yarns and of dyestuffs may be insufficient to meet the demand. The rising wages paid to the weavers and the higher prices of the materials pushes up the cost of the finished rugs until they become less competitive with those woven in nearby Afghanistan and Pakistan, as well as in other countries. Then too, many weavers contract for rugs, receive the proper amounts of yarn in advance, and then skimp on the required number of knots in order to finish the job with yarn left over which they may sell or weave into additional rugs.

Rugs are frequently woven on special order and for special occasions. To mark the celebrations of the 2,500th anniversary of the founding of the monarchy in Iran, held in 1971, the invited heads of state from many countries were asked to supply color photographs of themselves well in advance. At the celebrations they were presented with splendid silk rugs, about two by three feet, which displayed their portraits against the background of a Persepolis relief.

Lebanon, Switzerland, and Kuwait import large numbers of Persian rugs, but these countries are principally exporters. The most important sales are to the United States, the Federal Republic of Germany, and Great Britain. Current production of rugs is about four million square yards, equivalent to some 330,000 rugs of three by four yards each. These rugs are Iran's most valuable item of export, selling abroad for about $65 million each year.

Industrial Labor

The labor movement had its beginnings in Iran about 1921 and came into being at the hands of devotees trained by the communist party of the USSR. A number of guild unions and unions of government workers were set up at Tehran where the movement was headed by Sayyid Muhammad Dehqan who reported on his efforts at the fourth meeting of the Third International held in 1924. Several of his close associates were to reappear in the Tudeh party or in the autonomous movement in Azerbaijan in 1945. Thus, Ja'far Pishavari supported the unions through the paper *Haqiqat* published at Tehran, Ovanessian played an active role, and Reza Rusta came to Tehran from the Caspian area in 1927 to become one of the moving spirits. May Day celebrations were held at Tehran from 1925 through 1927 but in 1929 the government of Reza Shah closed down upon the movement, estimated to have 7,000 followers, and arrested some 50 of the leaders. In 1930 and later Rusta, Pishevari, Ovanessian and others were arrested and remained in prison or enforced residence until 1941 when political prisoners were released.

These leaders were not slow in renewing their activity. In 1942 the Tudeh party sponsored a Central Committee of Trade Unions which organized chapters in the main industrial centers of the country. The Central Committee made effective use of demonstrations, newspaper publicity, and strikes to gain higher wages for union members; Reza Rusta headed this activity.

In 1946 a cabinet decree authorized the implementation of

a comprehensive Labor Law, intended to fill the gap existing in this field in Iran. Some forty-seven articles regulated working hours, wages, holidays and paid vacations, and labor by women and children, as well as the terms of contracts between employers and employees, safeguards for the health of workers, an unemployment service, unions, boards of conciliation and arbitration, savings societies and social security. In this same year the newly created Ministry of Labor absorbed many of the established unions into the government-sponsored Union of Iranian Workers.

However, it was not until after the Tudeh party was banned in 1949 that Soviet and communist domination of the unions was broken: Rusta was arrested, surrendered bail, and fled the country. In 1951 the Iranian Trade Union Congress was founded by a consolidation of the Federation of Iranian Workers and the Central Federation of Trade Unions of Workers and Peasants. Other elements which joined included the Central Council of Unions of Workers and Farmers, the General Federation of the Unions of the Workers of Khorasan, and the United Front of Workers. Currently there are well over 300 unions, most of which are affiliated with the Iran Novin party. The Workers Organization of Iran appears to have supplanted the Iranian Trade Union Congress.

May Day is Labor Day in Iran and early in that month is held the National Labor Conference, which first met in 1969.

In January 1953 a law provided for workers' insurance and in August 1955 it was superseded by a more comprehensive Labor and Welfare Insurance Act, detailed in at least 75 articles. Concern for the condition of workers became especially marked after the White Revolution of 1963, and a number of programs were rapidly implemented or expanded.

The Labor and Social Welfare Institute is an independent organization authorized by the government to function in the fields of industrial relations, labor union affairs, management-labor affairs, cooperatives, industrial safety, and industrial research. Workers participate in over 500 Credit Unions, some 700 Workers' Cooperatives, and over 100 Housing Cooperatives. They receive benefits under the Social Insurance Act and the Social Security Act.

The Unemployment Insurance Act provides for payments

to workers who have become unemployed of minimum wages up to a maximum of six months. Capital for this fund comes from one percent of the worker's wages, and an additional one percent from the employer and from the state. State employment agencies are located throughout the country. Half those seeking jobs are between the ages of 18 and 29, and are unskilled or semiskilled. A nation-wide classification of positions has been carried out to assist both management and workers in getting the right people in jobs for which they are qualified. Strikes occur very rarely: disputes between management and workers are settled by arbitration councils which appear to favor the demands of the workers.

In 1972 the minimum daily wage for unskilled workers was between 80 and 100 rials, depending on living costs in regions of the country. Actually, most wages are above this minimum. Skilled craftsmen make about 500 rials daily. Vocational training is carried on in the factories, as are literacy programs, with those newly literate receiving increases in wages.

The Industrial Workers Profit Sharing Law was enacted in 1963 and subsequently amended. Workers at a factory or other enterprise sign group contracts with the management; they receive an additional one to three months pay each year as profit sharing, or, more precisely, in lieu of profit sharing.

Late in 1972 shares in state-owned factories were offered for sale to the workers and clerical personnel. This program is expanding to 140 state enterprises, with the general public given second priority on the purchase of shares. Local economists believe, perhaps somewhat naively, that when the worker considers himself as a co-owner of the factory he will work more assiduously and conscientiously, and this will result in higher production and a lower cost of the finished products of the factory.

Housing

Iran is very short of adequate housing, particularly for the low income groups. At Tehran speculation in land has pushed the price of desirable locations up to $150 a square yard, while rents have skyrocketed. Currently only 75,000 housing units are built throughout the country each year.

Agencies of the government have been active in this field, encouraging housing cooperatives and constructing housing for members of the armed forces, government employees, and low income workers. As an example of the latter category, south Tehran displays an extensive housing complex provided with all the services needed by this semi-independent community.

It has been estimated that during the period of the Fifth Plan some 2.2 million units will be needed, and the Plan Organization will construct 1,175,000 of these units. To discourage the influx of people into Tehran emphasis is being placed on housing in other urban centers. Major state enterprises, including NIOC and the large industrial plants, provide housing for their workers. Private industrialists are encouraged to follow this example, and are offered loans at low interest rates for the construction of approved types of housing. The Iran Mortgage Bank is the principal agent in this field, and has lowered its interest on housing loans to 6 percent. It is said that adequate housing can only be achieved if 5 percent of the GNP is channeled into this effort.

Foreign Trade

The foreign trade of Iran is a monopoly of the state. According to a bill passed in July 1953 the government has very broad powers in this field, including the right to establish import quotas on items which private concerns may bring into the country. The current pattern continues that in effect prior to World War II when measures of control introduced and conducted by the Iranian government enabled the country to maintain a rather precarious but favorable balance of trade. In years when the value of imports tended to exceed that of exports foreign exchange received from the operations of the oil concession was diverted to purchases abroad, and serious repercussions on the economic system of the country were avoided. This balance was upset during the war years and those which followed: unfavorable trade balances were influenced by the difficulty of reentering the competitive world markets and by the situation prevailing within Iran following the nationalization of the oil industry in 1951. Probably balances would have been even more unfavorable had not Iran executed a number of clearing agreements with European countries.

Prior to the beginning of each Persian calendar year the Ministry of Economy issues import-export regulations. These may include the total value of goods in certain categories which may be imported by merchants, may ban or restrict certain items of import and export trade, and may raise or lower customs duties on specific items. In general, the regulations are designed to protect local industries and to cut down on imports of luxury goods. These regulations incline to be very lengthy and complicated, and in cases where official consent is required to purchase certain goods the way is open for the exercise of influence.

The figures given in the following table, which represent the total value of Iran's foreign trade over a number of years, originated in publications of the Iranian customs administration. Converted from rials, these figures may or may not agree with other published statements. During the last decade the price of the rial in relation to foreign currencies has fluctuated as a number of rates have been available at the same time for various types of transactions. For the table conversion has been made at the rate of 32.5 rials to the American dollar up until 1955 when it was established at 75 rials to the dollar.

	Exports	Imports	Total Trade
1970	$277,900,000	$1,676,600,000	$1,954,500,000
1969	244,700,000	1,542,700,000	1,787,400,000
1968	216,900,000	1,389,200,000	1,606,100,000
1967	181,800,000	1,190,300,000	1,372,100,000
1966	157,500,000	963,700,000	1,121,200,000
1965	180,800,000	898,500,000	1,007,300,000
1964	153,100,000	757,100,000	910,200,000
1963	128,200,000	523,700,000	651,900,000
1962	114,700,000	558,700,000	673,400,000
1961	127,900,000	496,700,000	624,600,000
1960	111,300,000	702,100,000	813,400,000
1955	105,700,000	251,000,000	356,700,000
1950	107,500,000	189,100,000	296,600,000

The figures for imports include purchases of gold and silver but not the value of goods which were exempt from customs duties. The export figures do not include crude oil, refined products, and natural gas exported from Iran, nor the exports of the Iran Fisheries Company, not the value of trade conducted under barter agreements. An important obstacle against moving toward a more favorable balance of trade is represented by the amount of goods which come in duty free. As much as 40 percent by value of the imports are brought in by some twenty agencies of the government and charitable organizations. This number includes the Plan Organization, the NIOC, the Iranian State Railways, and the country's universities.

The major imports of Iran, in approximate order of descending value, are as follows: machinery; iron, cast iron, and steel; chemicals and pharmaceuticals; electrical machinery and appliances; silk, artificial silk, and artificial fabrics; automobile chassis and parts; paper, cardboard, and related products; wool and related products; rubber and rubber products; fats and fluid oils; vehicles and spare parts; tanning and dyeing extracts and dyestuffs; tractors; metals and mineral ores; wood and wood products; live animals and animal products; glassware and ceramic products; sugar; tea.

Major exports, also in descending order of value, are as follows: rugs; cotton; fresh and dried fruits; hides and leather; mineral ores; caviar; textiles and knitwear articles; soaps and detergents; gum tragacanth, caraway, and cumin seeds, medical and industrial herbs and seeds; shoes; sausage casings; manufactured goods; salt, sulfur, stone, lime, and cement; lambskins and other animal skins.

COMBINED EXPORTS AND IMPORTS OF IRAN'S
LEADING TRADING NATIONS[1]
(in millions of U.S. dollars)

	1970	1969	1968	1967	1966	1965	1964	1963	1962
USA	241.6	235.1	246.2	233.0	212.1	190.3	150.0	97.5	113.6
United Kingdom	172.5	197.5	189.4	149.1	131.8	129.3	114.3	91.0	115.0
Japan	217.8	175.7	139.0	95.3	77.9	75.4	51.4	41.2	31.6
West Germany	387.1	342.7	328.4	302.7	228.5	206.3	162.9	122.3	115.1
USSR[2]	193.6	173.4	85.1	63.3	47.1	34.0	45.0	41.4	31.8

[1] Excludes exports of crude oil, refined products, and natural gas.
[2] Conducted by means of clearing agreements.

The imbalance of Iran's trade is reflected in the figures for 1970: United States; exports to Iran $217.4 million, imports $24.2 million: Japan; exports $201.2 million, imports $16.6 million: West Germany; exports $347.9 million, imports $39.2 million. As mentioned in a previous section, Iran has concluded a number of agreements with the Eastern European countries, Japan, and China which open markets for the rising output of the country's industrial and manufactured products. The states of the Arabian littoral of the Persian Gulf represent an expanding market for Iranian goods.

Banking

Before the opening of the twentieth century Iran had no adequate system of government finances. The national revenues had always been the personal property of the rulers, who made their own decisions as to what amounts they would pass on for the upkeep of the army or to their personal favorites. Because ready cash was always wanting, loans were obtained from European powers under terms which assigned customs revenues or granted concessions to the creditor nation making the loans.

As soon as the Constitution was in force, leaders of the country realized that foreign specialists must be called in to help in the creation of a stable financial system. At first Belgians and French were employed, and in 1911 came an American, W. Morgan Shuster, whose disinterested devotion won the hearts of the Iranians before he was forced out by foreign pressure on the Iranian government. Chaos reigned until the period of Reza Shah. Then a request was again made for American advice, and in 1922 A. C. Millspaugh arrived, at the head of a mission which in the next five years succeeded in establishing a balanced budget and in assuring the efficient collection of fair taxes. After the departure of the mission in 1927 the government maintained a balanced budget for several years, but was finally compelled to take over the control of trade and industry in an attempt to cope with the problems of the general world depression.

Mr. Millspaugh returned to Iran by government invitation in 1943, and requested and was granted very broad powers.

His staff of Americans, numbering as many as fifty at one time, made a serious attempt to reform the entire financial structure.

The comprehensive program of the Millspaugh Mission was complicated and slowed by the emergency conditions of the war, which also made it difficult for the Iranians to observe what progress was being made. Parliament began to limit the powers of the Mission, and as a result Millspaugh resigned and left Iran in 1945. His experiences during these two hectic missions are described in books published in 1925 and 1946.

Long the most important bank in the country, the Imperial Bank of Persia (later Iran), was founded near the end of the nineteenth century, operating under a sixty-year concession. In 1949 the concession ended and following negotiations with the Iranian government its directors agreed to carry on business under new, somewhat stringent regulations, and under the name of the British Bank of Iran and the Middle East. These regulations proved too restrictive and in 1952 the bank withdrew from Iran. This left the field of foreign banking to the long active Soviet-owned Russian Bank in Iran which operated under these same regulations and which employed its capital of 100,000,000 rials to promote trade between Iran and the USSR.

The Iranian constitution had envisaged the creation of a state bank, but it was not until 1927 that the *Bank Melli Iran*, or the National Bank of Iran, was established. It grew very rapidly and in 1932 took over the privilege of note issue from the Imperial Bank.

A Monetary and Banking Act was passed in 1960. Prior to that time the Bank Melli had found it difficult to operate both as a commercial bank and the state bank, and this act provided for its reorganization. As the Bank Melli it is the country's largest commercial bank, with 1,200 branches and offices in a number of foreign countries. Split off from it was the new *Bank Markazi Iran*, the Central Bank of Iran, which took over the handling of government accounts, the issue of bank notes, and the enforcement of monetary laws and regulations.

The Central Bank maintains full coverage of the bank notes

in circulation by backing them 30.49 percent by gold, 9.51 percent by foreign exchange and shares in such institutions as the International Monetary Fund and the International Bank for Reconstruction and Development, and 60 percent by government obligations secured by the value of the crown jewels.

Other state-owned banks are the *Bank Sepah*, Army Bank; *Bank Keshavarz: Iran*, Agricultural Bank; *Bank Rahni Iran*, Iran Mortgage Bank; *Bank Etebarat Sanati*, Industrial Credits Bank; *Bank Bimeh Iran*, Iran Insurance Bank; and the *Bank Refah Kargaran*, Workers' Welfare Bank. All the state-owned banks had paid up capitals of 24 billion rials. There are ten private domestic banks, and ten banks in which foreign capital participation ranges from 20 percent to 49 percent. The Bank of America and the First National City Bank of New York maintain offices in Tehran. In 1971 the total assets of all the banks operating in Iran was 810 billion rials.

Public Finance

For several years the mounting costs of the regular expenditures of the government, including those of the armed forces, and of development projects in which the emphasis is on the rapid industrialization of the country have resulted in increasing annual deficits. These deficits are made up by loans of banks to the government, by floating issues of government bonds, and by negotiating foreign loans. The accompanying chart illustrates the situation.

Electric Power

An orderly and unprecedented expansion of the country's electric power systems took place after 1964 when the Ministry of Water and Power was established and an Iran Electrical Authority was formed. A number of regional companies were brought into being: the Tehran Regional Electric Company, the Khorasan Regional Electric Company, the Western Regional Electric Company, the Sefidrud Regional Electric Company, and several others. Each company constructed a regional grid which serves a number of cities, towns, and larger villages.

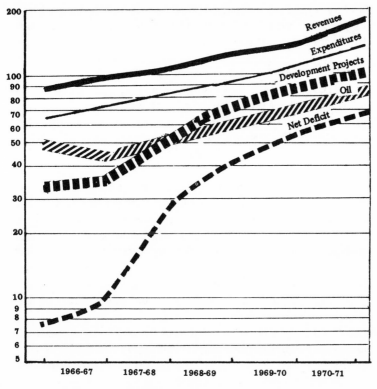

| | 1966-67 | 1967-68 | 1968-69 | 1969-70 | 1970-71 |

Source: Central Bank of Iran

As noted earlier, major dams supply much of the total power, with Tehran receiving a part of its requirements from power stations at the two dams in Khuzistan. Thermal power stations utilize coal, diesel oil, and natural gas, with the construction of natural gas stations on the increase. Plans for a number of atomic power stations have been under study, but the real necessity for such installations seems open to question.

Although the available statistics are not in agreement, it would appear that in 1973 some 5.8 billion kilowatt hours of electricity were generated, of which a total of about one-third was consumed by industrial enterprises. In the near future this figure is to rise to 12 billion kilowatt hours, with some two-thirds used by industries.

Telecommunications

In 1972 the Iran Telephone Company, a state-owned organization which took over from a private company in 1950, was replaced by the Iran Communications Company, also state-owned. This action followed the conclusion of a $175 million contract with an international consortium to install a microwave network providing direct dialing of calls throughout the country. Automatic exchanges serve most of the country with 225,000 of the country's 411,000 phones at Tehran.

The Regional Cooperation Development organization maintains a microwave telecommunications system which links Ankara, Tehran, and Karachi. A telex system connects Tehran with Germany, France, and Great Britain, and there are telephoto connections with the countries just named and with Japan, Italy, and India. A Telecommunications Training Institute functions at Tehran.

Since 1969 an earth satellite station located at Assadabad, near Hamadan, provides telex and telegraph facilities, as well as the reception of live television programs from abroad for rebroadcast in Iran.

Roads

The history of roads and trade and travel in Iran goes back into remote antiquity. In the Achaemenid period the famous "Royal Road" ran from the city of Susa through Mesopotamia and Asia Minor to the city of Sardis, just inland from the Aegean coast, for a total length of about fifteen hundred miles. The post system introduced in this period and carried on in one form or another into the twentieth century is based on a series of rest houses, located at convenient intervals, where official messengers or travelers can obtain lodging, food, and fresh horses.

By the beginning of the Christian era the great east-west highway from China to the Mediterranean was heavily traveled, and was known as the "silk route." Over it not only silk but quantities of pottery, spices, and other wares were transported from the Far East to the markets of the western world. The camel caravan was the main method of transport through many long centuries until a network of trade routes grew and covered Iran. The lines of some of these routes are paralleled by modern roads, while others are marked only by deserted caravanserais and ruined villages. The security of caravans traveling them was always a major consideration; for example, in the early fourteenth century guards were stationed along all important routes, and local officials were compelled to make good any losses caused by attacks on caravans passing through their districts.

Caravanserais may be seen at intervals of about twenty miles along every road. Many were erected by Shah 'Abbas in the early seventeenth century; according to legend the number reached nine hundred and ninety-nine. Others are far older than his time, while those of quite recent date were built by local governors, by pious wealthy men, or by the innkeepers themselves. Constructed of baked brick or stone, they are rectangular in plan with a large, open central court and a well in the center, and around the court are rooms for the men and stables for the beasts. Welcome as the sight of a caravanserai must have been at the end of the day, one early traveler was not too well satisfied with them, for he wrote: "Coming to our Inns, we have no Host, or Young Damosels to bid us welcome, nor other furniture than bare walls; only an open house with no enlivening glass of wine to encourage the badness of the march, but after the fragments of yesterday's provisions we betake ourselves to rest with much eagerness amid the noise of carrier bells, feeding, neighing, braying and with the singing, chatting and din of the servants and muleteers."

Across the level stretches these early routes were mere tracks, but in the mountain passes wide, ladder-like steps were cut out of the bare rock. Shah 'Abbas fostered a better type of road, called the *sang-i-farsh* or "stone carpet"—a stone pavement laid upon a high embankment of earth, sections of which may still be seen south of Tehran and along the Caspian shore. Today the camel caravans are less frequently seen along

the many motor roads, but they are still common in the eastern part of the country where the age-old trails cross desert
wastes impassable to motor traffic.

The camel himself is one of the major curiosities of nature;
as an old Arab story has it, God created him last of all the
animals, out of all the ill-matching odds and ends that were
left over. An observant European who traveled in Iran some
two centuries ago has this to say: "Camels, a beast abounding
in Persia, and of great use, esteem and value in those Oriental
parts; long-lived they are, oft-times exceeding three score
years, of disposition very gentle, patient in travel, and of
great strength, well enduring a burden of towards a 1000
pound weight; content with little food and that of the meanest sort, as tops of trees, thistles, weeds and the like, and less
drink, in those dry countries, usually abstaining little less than
four days."

When setting out on a long journey the caravans usually
get under way in the late afternoon and travel only about
three miles, so that after the first camp has been made and the
equipment checked, someone can be sent back to collect those
items which have been left behind. Road distances are meas-
ured in *farsakhs*, which are not always of uniform length: a
farsakh on level ground is just under four miles while in hilly
country it is about three miles. The caravan covers about half
a farsakh an hour; an average day's travel is between four
and five farsakhs.

In hot weather or on very long trips, the caravan starts
after sunset and travels all night. The lead camel and the
one at the end of the long procession carry great bells hung
to their necks, and as long as the bells sound the line keeps
moving. Each stage of the journey is marked by a caravanserai, where the loads are removed and the camels sent out to
graze during the day while the men sleep.

Throughout the country the single-humped Arabian camel,
or dromedary, predominates, while toward the northeast the
two-hump Bactrian camel is much more frequently found.

More modern roads became a necessity after the introduction, in the nineteenth century, of wheeled vehicles. The first
modern road built may have been that from Tehran to Qum,

constructed in 1883. By 1899 a new road was open from Rasht, on the Caspian, to Tehran, one section of which was built by a Russian company and the rest by the Iranian government. Other early roads connecting Russia with points in northern Iran were usually treated as concessions, and the builders permitted to charge for the use of the roads and for the horses and lodging supplied along the way. The concessions were held by a Russian company and a British company called the Iranian Road and Transport Company. By 1914 there was a network of these carriage roads, one branch of which extended to the Persian Gulf from Tehran. But motor traffic was still very limited, and when in 1919 a British task force traveling in Ford cars attempted to reach Qazvin from Baghdad they were forced to do a great deal of engineering work along the highway.

During the reign of Reza Shah a sustained effort was made to improve the road system. The early roads were neither wide enough nor built with a solid enough base to sustain heavy traffic, and they were entirely rebuilt of gravel on a base of crushed stone and provided with hundreds of bridges.

Before 1941 stretches of the three main routes leading out of Tehran had been asphalted. During the war, when the constant pounding of the truck convoys carrying supplies from the Persian Gulf to Russia made necessary a more permanent and resistant surface than gravel, the British, American, and Russian engineering units either executed or supervised the asphalting of long stretches of the supply route. At the end of the war the road from Khanaqin, on the frontier of Iraq, to Qazvin; the long route from Khorramshahr through Ahwaz, Khorramabad, and Malayer to Hamadan; and a part of the road from Rasht, on the Caspian Sea, to Qazvin had been asphalted—a total length of just over nine hundred miles. Another long stretch of asphalt runs from Tehran to Baghdad, and then on across the Syrian desert to Damascus and Beirut.

The road network has been greatly expanded in recent years. The main highways have been rerouted in large part, with modern engineering design doing away with the steep slopes and sharp curves of the earlier alignments. In 1973 there were 9,000 miles of asphalted highways and 8,500 miles of gravel surface roads. Also, 12,000 miles of gravel and dirt feeder roads which linked rural areas to each other and to urban centers, thus increasingly facilitating the marketing of agricultural products. There are no concrete paved highways: asphalt is supplied by the petroleum industry, while cement is more expensive to transport and in long stretches of the country the water essential for mixing concrete is lacking.

Major highways across the country have been integrated into regional development plans. CENTO has been responsible for the highway which runs from the Turkish frontier to Pakistan. Two sections of the Asian Highways sponsored by the Economic Commission for Asia and the Far East (ECAFE), a United Nations agency, have been built. The stretch of Asian Highway No. 1 runs from a different point on the Turkish frontier to Tehran, then north to the Caspian Sea and along northeastern Iran to Mashhad and on to Afghanistan. Asian Highway No. 2 runs from the Iraq frontier to Pakistan. The routes of these and other main roads may be followed on the map of Iran.

Iran has some 325,000 automobiles, including 50,000 taxis, 18,500 buses, and 55,000 trucks. Over half the automobiles are in Tehran and in the morning and afternoon massive traffic jams develop on the main roads of the city.

Railways

Throughout the nineteenth and early twentieth centuries both Great Britain and Russia were interested in the construction of rail lines through Iran as a means of furthering the economic and commercial interests of the two powers. The line proposed by the British would have crossed southern Mesopotamia, south and southeast Iran, and Baluchistan to link India and the Mediterranean. Russia planned a line across the northern breadth of Iran which would link the Caucasus with Turkestan, and would also serve to strengthen Russian economic penetration in her chosen sphere of influence.

In the end neither country was able to carry through its plans, and they constructed only a few short lines of very little value to Iran. The first was a narrow-gauge line built in

1892 to connect Tehran with the shrine village of Shah Abdul Azim, some six miles to the south. In 1914 a Russian company began the construction of a ninety-mile line from Julfa to Tabriz, which, together with a thirty-mile branch to the shore of Lake Rezaieh, was completed in 1916. The track is of the Russian broad gauge and connects directly with a main Russian line by means of a bridge over the Aras River on the frontier. Ownership of this line came to the Iranian government through one of the clauses in the Irano-Soviet Treaty of 1921. A short narrow-gauge rail line once ran from the town of Rasht to the shore of the Caspian, and a narrow-gauge line was built by the Anglo-Iranian Oil Company in 1923 for the movement of supplies and equipment within the area of the oil fields. During the first World War the government of India extended its Trans-Baluchistan railway from Mirjaveh, on the Iran-Baluchistan frontier, to Zahidan, some fifty-two miles inside Iran. The line was the very wide Indian 5'6" gauge, and after the war the tracks inside Iran were taken up. After 1941 they were relaid as standard-gauge, and the line is now in service from Zahidan to Pakistan.

Even before Reza Shah became ruler of the country, steps had been taken toward the building of a Trans-Iranian railway which would serve less as a mere link with railroads in adjacent countries than as a means of promoting Iran's own interests. The route selected, from the Persian Gulf to the Caspian, offered several advantages. It would enable the central government to maintain closer liaison with the provinces and provide for rapid troop movements necessary to maintain internal security. It would provide for a better distribution of the country's natural resources; food products of the fertile Caspian littoral could easily be transported to the arid south, and minerals and raw materials could reach the newly established industrial enterprises. It would also, by establishing an Iranian port of access on the Gulf, help to reduce Iran's dependence upon Great Britain and Russia for her commerce.

As early as 1925 Parliament enacted a law which gave the government a monopoly on sugar and tea and a bill providing for a tax on the transport of merchandise by road: funds from these sources were allocated exclusively to railway construction. A bill providing for survey work was passed in 1926, and in 1927 another law authorized construction of the railway from the Persian Gulf to the Caspian Sea. In 1928 a syndicate of American and German engineering firms was awarded a contract for the survey of the entire line and the construction of sections south from the Caspian terminus and north from the Persian Gulf, at a fee of 10 percent of the costs. In 1931 this contract was canceled by the Iranian government, and the job was turned over in 1933 to Kampsax, a syndicate of Swedish and Danish firms, who proceeded to let various stretches of the line to local and foreign contractors.

In 1938 the single track, standard-gauge line was in operation from the newly built port of Bandar Shapur on the Gulf through Tehran to the new port of Bandar Shah at the southeast corner of the Caspian, for a total distance of 865 miles.

Probably the construction of no other railway line in the world has met with more natural difficulties than did that of the Trans-Iranian, much of which is through very mountainous country. The southern section, from the Persian Gulf to Tehran, runs for miles on ledges blasted out of the precipitous walls of deep gorges, and finally climbs to the plateau through a pass 7,253 feet above sea level. Hundreds of bridges had to be built, and in this section there are 125 tunnels with a total length of 35 miles. The northern section makes the ascent from the Caspian shore to the plateau in a much shorter distance, with a gradient as high as one in thirty-six. The final pass is traversed by a tunnel more than two miles long, at an altitude of 6,924 feet. In this section, on which forty to fifty thousand workmen were employed, the length of the 65 tunnels totals 12 miles, and in one stretch the line winds for 33 miles to cover a distance of only 18 miles as the crow flies.

Stations, warehouses, sidings, and water tanks were erected along the entire line at intervals of not more than twenty miles. Many special problems, such as the necessity for water-softening equipment in many areas, were met and solved. Rails, locomotives, cars, structural materials, and machinery were all brought in from abroad; only the wooden ties were domestically produced. The final cost of this line has been estimated at between $150,000,000 and $200,000,000, all of which was raised within the country and represented a serious drain on the national income.

Railway construction did not end with the completion of the main Trans-Iranian line, for work had already been started on two other long lines which would connect eventually with systems already in operation in neighboring countries. During the war the operation of the Iranian railway was taken over by the Allies in order to expedite the movement of supplies to Russia. The Iranian government was to receive payment for the use of the system according to the volume of freight carried and England paid some £20,000,000. British, American, and Russian forces were responsible for the operation of different sections of the line; scores of locomotives were shipped far across dangerous seas into the Persian Gulf to be brought into the country, and hundreds of cars and several million tons of supplies were carried over the line. The Americans and British did some additional construction work, the British building a new line from Ahwaz to Khorramshahr, on the Persian Gulf, and a spur from this line to Basra in Iraq.

In 1973 some 3,000 miles of rail lines were in operation. Major additions to the earlier network include a line from Tehran to Mashhad, completed in 1957, and one from Tehran to Tabriz, completed in 1958. In 1971 a line was finished from Tabriz to the Turkish frontier to connect with a Turkish railway. Passengers are now carried from Iran to Istanbul and on to Europe by this line, crossing Lake Van in Turkey by ferry. The line south from Tehran to Kerman sends off a branch to Isfahan, while another line from the Arya Mehr steel mill near Isfahan, mentioned earlier, connects with the main line to Kerman. From Kerman this line is being extended to Zahidan to connect with a Pakistan link which continues on to Karachi. A line from Kerman to Bandar 'Abbas is under construction.

The rail network carries some 3.5 million passengers annually and more than 50 million tons of goods, exclusive of the materials transported to the Arya Mehr steel mill. It operates at a loss, spending 11 rials for every 4 rials of income.

Air Services

Iranian Airways, a privately-owned company, began operations in 1946 and later entered into a management contract with Pan-American International Airways Corporation. In 1962 it was taken over by Iran Air (Iranian National Airways Corporation), a newly formed state enterprise. It is sometimes known as Homa after its symbol, a mythical bird of ancient Iranian legend. In addition to a number of older types, it operates Boeing 707s, 727s, and 737s. An order has been placed for a Concorde. It has a capital of 4 billion rials.

Iran Air has frequent flights to nearly twenty cities and towns within the country. It serves the states of the Arabian littoral of the Persian Gulf with flights to Bahrain, Dhahran, Dubai, and Abu Dhabi. Its international service to Europe lands at Moscow, London, and several other capitals. To the east it flys to Kabul, Karachi, and Bombay, and in 1973 was to extend this service to Peking. Each year the airline carries as many as 45,000 Muslim pilgrims destined for Mecca, landing them at nearby Jidda. Carrying well over 700,000 passengers a year, with a load factor of about 55 percent, the airline operates at a profit.

In 1970 a second airline, Parsair, began operations with small planes which service the less important provincial towns. The Air Taxi Company operates out of Tehran, flying to chosen destinations.

There are international airports at Tehran, Shiraz, Abadan, and Bandar 'Abbas. Mehrabad airport, some six miles west of Tehran, is served by some twenty foreign international airlines. A new airport is under construction about 20 miles southeast of the capital: with four times the capacity of Mehrabad; it is expected to cost $120 million.

The air services come under the control of the Department of Civil Aviation which maintains the Higher Institute of Civil Aviation Studies.

Ports and Navigation

Passengers and freight on the Caspian Sea are carried in Soviet ships. Iranian ports on the Caspian coast are Bandar

Pahlavi, Noshahr, Mashhad-i-Sar, Bandar Gaz, and Bandar Shah, of which Bandar Pahlavi and Noshahr are the most important. These last two named handle some 400,000 tons of cargo a year which moves between Iran, the USSR, and the countries of Eastern Europe by means of a joint Irano-Soviet transport company. From the northern end of the Caspian cargo moves via the Don-Volga and Don-Baltic canals to reach either the Mediterranean or the Baltic Sea.

The principal ports along the Persian Gulf are Khorramshahr, Bandar 'Abbas, Abadan, Bandar Shapur, and Bandar Mashur. The last three are outlets for the petroleum and petrochemical industries, as mentioned earlier, as is the island of Kharg. Khorramshahr, the principal port of the country, lies just upstream from Abadan on the Shatt al-'Arab. Its capacity is about two million tons a year. A new port has been constructed at Bandar 'Abbas which handles general cargo and up to 900,000 tons of chromite ore for export. Less important ports along the Persian Gulf include Bushire, Bandar Lingeh, Jask, and Chah Bahar, but each of these has been provided with new facilities in keeping with the overall plan for the development of the Gulf. The principal ports named are served by a score of foreign shipping lines. In addition, the local Arya National Shipping Line, with eleven vessels, maintains a circular route around the Persian Gulf and sends its ships to Atlantic ports and to the Pacific, including Shanghai.

Navigation of Lake Rezaieh in the northwest of Iran is limited to small steamers and barges which make a circuit of the lake, calling at several points along its shores. Since the level of this large body of salt water is slowly sinking, this traffic may decrease. Of the many rivers of Iran only the Karun, which discharges into the Persian Gulf near Khorramshahr, is navigable for any distance. Barges and launches ply this river between the Persian Gulf and Ahwaz.

IX. FROM MONARCHY
TO ISLAMIC REPUBLIC

I N the concluding chapter, "Trends Toward Tomorrow," of the eighth (1976) edition of this work it was stated that Muhammad Reza Shah Pahlavi was very much in command of his nation. He insisted that the only political opposition came from terrorists, from communists guided from without, and from extremely conservative Muslim elements. No experienced observer of the Iranian scene would have believed that a single elderly, frail, extremely conservative Muslim could topple the monarchial system which had held sway in Iran for over two thousand years.

Ayatullah (Sign of Allah) Ruhullah Musavi Khomeini was born in the village of Khomein in 1902. At the age of fifteen he went to Iraq for religious studies and in 1922 settled in Qumm where he became a very popular teacher at one of the religious schools. In 1941 he published the *Kashf al-Asrar*, "Uncovering of Secrets." He asserted that until the return of the Imam, learned religious figures are his representatives. Civil authority is usurped and invalid: to grasp this concept readers are referred to the pages of this present work which describe the Shi'a sect and its belief in the Hidden Imam.

In 1961 the Ayatullah Borujerdi, the leading religious figure and one who was on fairly good terms with the government, died. Khomeini was now freer to extend his attack on the civil state to include all aspects of the modernization of Iran which aped Western models of social reform. Reforms had seriously weakened the position of the clerics in areas where they had been strong. The reforms included civil and penal codes and courts based on Western models; registration of documents in civil offices; a uniform dress law that limited the wearing of clerical garb; bringing women into society through granting the franchise, removing the veil, and introducing coeducation; land reform; and a secular educational system. Khomeini's denunciation of the influence of the West coincided with the spirit of xenophobia, endemic in Iran because of prolonged foreign interference.

In addition, there existed concern among the intellectuals over the destructive impact of Western manners and modes on Iranian culture: the reader is referred to page 210 for expressions of this concern. Thus, the Ayatullah launched a two-pronged appeal which reached a larger number than one limited to Islam would have.

Earlier pages describe the six reform measures, sponsored by the Shah and called by him the White Revolution, which were approved by a popular referendum in January 1963. Khomeini opposed the reforms, calling land reform, which distributed farm land to peasants, a fraud. In April he spoke at Qumm against the government and in June attacked the Pahlavi dynasty and Israel, insisting that the religious leaders were not reactionary. Khomeini was immediately arrested, three days of rioting at Qumm followed, resulting in a number of deaths and the imposition of martial law. In 1964 he was released to Qumm before his exile in November. Just at this time he was able to exploit a popular feeling against the regime when the Parliament passed a Status of Forces law which conveyed a measure of diplomatic immunity to a number of Americans employed in Iran and their families. Fanning the flames of xenophobia, his denunciation of the West focused directly on the United States, called Amrika in Iran.

In November 1965 the scene of exile shifted from Turkey to Iraq. It was there that he published his *Valiyat-i Faqieh*, which translates as "Rule of the Religious Jurist," with a subtitle equivalent to "Islamic Government." Although not so stated, Khomeini himself was to be the *faqih*, or jurist, and in this respect the successor of 'Ali, the hero of Shi'ism, one of whose titles was "Shah of the State." The text reiterated his claim that the mujtahids, the highest level of the Muslim jurists, must rule, that the monarchy must be abolished, and that the faithful must take action to achieve these ends. Apologists have questioned the authenticity of certain passages which do appear in the text, such as this one (my translation):

Before us we see the Jews making a mockery of the Qoran. It is our duty to reveal this treachery and to shout at the top of our voices until people understand that the Jews and their foreign masters are plotting against Islam and are preparing the way for the Jews to rule over the entire planet. I greatly fear that, by their own special methods, they will indeed realize their desired aims. In Tehran, Christian, Zionist, and Bahai missionary centers issue their publications in order to mislead people and to alienate them from the teachings and principles of religion. Is it not our duty to demolish these centers?

On 7 January 1978, the Tehran newspaper *Ettela'at* printed a government release that contained a scurrilous attack on the background, motives, and ambitions of the Ayatullah Khomeini. The religious students at Qumm protested the article; the atmosphere became tense and on the 9th rioting broke out, with casualties. Demonstrations took place in other towns—at Isfahan, at Shiraz, at Mashhad, where rioting reached huge proportions in July, and at Tabriz in August. In Tehran, thousands came out in the streets on 7 September and on Friday, the 8th, troops fired into the crowds, killing many people. This date became known as Black Friday and was a rallying cry against the regime. Here space does not permit tracing the course of demonstrations in any detail.

Several of the important elements who took part in the mass demonstrations have been identified:

1) The Muslim clerics. Statements made by Khomeini and others were relayed throughout the country by the numerous mullahs, the lowest level of the clergy.

2) The bazaari. At Tehran these were the wholesalers, retailers, craftsmen, private bankers, and laborers from the vast bazaar and from the bazaars elsewhere. The bazaaris were traditionally tied to the clerics through financial contributions and their devotion to the "old-time religion."

3) The well-educated who were eager to take part in movements for Islamic reform. Among those advocating reforms were Mehdi Bazargan and Dr. 'Ali Shariati. Shariati was born in 1933, a son of an ayatullah. While studying at Mashhad, he became embroiled with the regime and was jailed for some months. In 1959 he went to France and spent five years at the University of Paris, studying sociology, writing about Islam,

and engaging in anti-imperialist activities. In 1964 he returned to Iran where he was arrested, released, and then permitted to lecture for some time at a religious center at Tehran. Again arrested, on his release he went to England where he died suddenly in June 1977. A number of his lectures and articles are available in a slim volume entitled, in translation, *On the Sociology of Islam*. Opposed to the extreme fundamentalists and to the rootless intellectuals, he denounced what he saw in Iran as the "nauseating apish imitations of Europe."

4) The so-called, in sociological language, in-migrants who had flocked from the villages to Tehran and who lived in depressed conditions. By 1976 they numbered 1.4 million, and a great many were only marginally employed.

5) A mass of young people. These included high school graduates without employment and graduates denied admission to the universities because applicants outnumbered places by many times: most of the martyrs were from this group.

Disciplined groups of leftists included members of the Tudeh Party and the Fedayan-i Khalq. It is possible that they directed the demonstrators to burn scores of branch banks, cinemas, airline offices, and department stores, with a view to paralyzing public activity, while the Mujaheddin-i Khalq destroyed the liquor stores.

The Shah and his advisers sought ways to pacify the protesters by showing a willingness to grant political freedoms. Near the end of August 1978 Jafar Sharif-Emami, an elderly statesman alleged to have some ties to religious leaders, was named prime minister. The concessions he offered were ignored by the demonstrators and in October a series of strikes took place, with the results detailed elsewhere in these pages. On 6 November, the Shah named General Gholam Reza Azhari to head a cabinet of ten ministers, of whom six were military men. He vowed to restore peace and order as a precondition to holding free elections; his efforts failed as between 26 November and the announcement of martial law on 2 December troops and tanks battled crowds.

Early in November 1978 the Ayatullah Khomeini moved from Nejaf in Iraq to Paris where the cynically benign attitude of the French authorities enabled him to meet with associates and admirers, to make phone calls to Iran, and to record on cassettes which were copied in large numbers and sent into Iran. Before he came to Paris, associates had published savage attacks against imperialism and its support of the Shah, and the French press echoed these sentiments. It could be suggested that the French government thought the fall of the Pahlavi regime would be in its best interests.

The role of the United States in the downfall of the regime has been the subject of virulent and self-serving articles. Full-length works on the history of relations between the two nations are in preparation. In briefest terms, it appears certain that the Carter regime was very late in realizing the seriousness of the challenge to the Shah and made little effort to consult with those many Americans who had an intimate knowledge of Iran. In December former Under Secretary of State George Ball, whom the Shah labelled as "no friend," was asked to appraise the situation: his word was, in effect, that the Shah must go. Early in January General Robert Huyser, deputy commander of United States Forces in Europe, arrived unannounced at Tehran. He met with the American ambassador and once with the Shah and the ambassador: their message was to advise the Shah to leave Iran.

Published accounts of the Shah's last weeks in Iran picture him as unsure of himself and as vacillating: most accounts failed to indicate that he was already a very sick man, too sick to be able to cope effectively in the absence of the support from the United States that he had every reason to expect. American politicians may have learned from the Iranian experience that pressing friendly countries too hard for human rights concessions when these governments are under attack by internal and external enemies may be inimical to the foreign policy objectives of the United States.

The Shah and the Shahbanu left Tehran on 16 January 1979 for a rather brief life in exile until his death at Cairo on 27 July 1980. About ten days before his departure a new prime minister, Shapur Bakhtiar, had presented a cabinet and on the departure of the Shah shared power with a nine-person Regency Council. On 1 February 1979, Ayatullah Khomeini flew to Tehran on an Air France plane and set up a rival government. By the middle of February the position of Bakhtiar had become intolerable and he left Iran. Khomeini named a Revolutionary Council to run the country. Neither the names nor the number of its members were disclosed, although over a period of time the names of some fifteen members were revealed, of whom the majority were religious figures. Its offensive arm was the Pasdaran, well translated as the Islamic Revolutionary Guards: it comprised some 60,000 armed men. Another arm was local Komitehs, the Persian adoption of "committee," which tried, convicted, and executed persons or turned them over to the courts of the Revolutionary Council.

The Revolutionary Council prepared to draft a new constitution. The existing Constitution of 1906, modified and amended at later dates, is described in Chapter VI. It provided for both an Assembly, or Majlis, and a Senate. One provision recognized that the sect of the Twelve Imams was the official religion of the country, and an article stated that five or more ulemas chosen by the Majlis were empowered to reject any proposed legislation that was not in accord with the sacred principles of Islam. This article was not to be altered in any way until the appearance of the Imam of the Time, but it was never implemented.

The referendum held on 1 April 1979 was overwhelmingly in favor of the establishment of an Islamic Republic of Iran. The Ayatullah announced that a Constituent Assembly would be chosen to draft a constitution. Instead, an Assembly of Experts was elected in the opening days of August 1979 and entrusted with the task. There were to be some seventy-five such experts, but information is available only on the election of ten members at Tehran. The Ayatullah Taleghani received over 1.7 million votes, followed by Bani Sadr, the Ayatullah Montazeri, Dr. Ali Golzadeh Ghafuri, Sayyid Beheshti, and others. In November the Assembly of Experts completed its work and the draft was approved by a referendum held early in December 1979.

The Constitution of the Islamic Republic of Iran is a very lengthy document. A preamble of about 4,000 words reviews the revolution and discusses the ideals of an Islamic state. The document itself contains 175 Principles, many of which are quite long. In this respect the constitution is in contrast with others recently adopted elsewhere, where the approach has been to outline general principles which are to be defined in detail by legislation.

Key features of the Constitution appear in several of its Principles. Principle 5: "During the absence of the Glorious Lord of the Age (the Twelfth Imam), may Allah grant him relief, he will be represented in the Islamic Republic of Iran as religious leader and imam of the people by an honest, virtuous, well-informed, courageous, efficient administrator and religious jurist, enjoying the confidence of the majority of the people as leader. Should there be no jurist endowed with such qualifications, enjoying the confidence of the majority of the people, his role will be undertaken by a leader or council of leaders, consisting of religious jurists meeting the requirements mentioned above, according to Principle 107." The name of this religious jurist is given in Principle 1. "The Great Ayatullah Imam Khomeini."

Principle 12: "The official religion of Iran is Islam, and the sect followed is Ithna Ashari (Twelve Imams) . . ." Principle 56: "The absolute ruler of the world and of humanity is Allah . . ." Principle 107: "Whenever one of the jurists who fulfills the conditions mentioned in Principle 5 of the law is recognized by a decisive majority of the people for leadership and has been accepted—as is the case with the Great Ayatullah Imam Khomeini's high calling to the leadership of the revolution—then this leader shall have charge of governing and all the responsibilities arising from it . . ."

Principle 110 deals at length with the duties and powers of the leadership, as reflected in these excerpts from it: the leader appoints the jurists on the Council of Guardians, names the highest judicial authorities, and exercises command over the armed forces; including the power to declare war. After the election of a president, he signs an order naming him to the post, and may, under certain circumstances, dismiss him.

The president is elected for a four-year term by direct popular vote and has broad executive powers. He chooses the prime minister, with that choice being approved by the Assembly. The National Consultative Assembly has 270 members elected for a four-year term. The Council of Guardians consists of six just and religious persons and six lawyers in different branches of the law, all elected for a term of six years. This council examines all proposed laws and regulations to insure that they conform to Islamic standards and the principles of the Constitution.

A number of the Principles appear to reflect Islamic idealism. For example, Principle 31 reads: "A suitable dwelling, according to need, is the right of every Iranian person and family. The government is responsible for providing this, on a priority basis, to those who need it most, in particular to the peasants and agricultural workers." Some Principles may already have been violated by actions of the government. These statements include, "Unarmed assemblies and marches are permitted provided that they do not violate the precepts of Islam," and "Publications of the press are free in the expression of topics unless it is contrary to Islamic precepts or public rights."

On 25 January 1980 Abul Hasan Bani Sadr was elected president by over 10 million of the 14 million votes cast: he was then forty-six years of age. Exercising his executive powers he named, although reluctantly, Muhammad Ali Rajai as prime minister, but rejected a number of the names of his proposed cabinet. A son of the Ayatullah Sayyid Nasrullah Bani Sadr of Hamadan, he attended secondary school at Tehran and went on to the University of Tehran where he studied theology and majored in economics. Arrested in 1961 and 1964 for antigovernment activities, he went to Paris for further studies. He wrote articles related to Islamic ideology and the economy of Iran, and was coauthor of *Pétrole et Violence*, critical of the oil policies of Iran. At Paris he was a member of the Committee for the Defense and Promotion of Human Rights in Iran.

Returning to Iran early in 1979, he was a member of the Assembly of Experts, later minister of foreign affairs, and then minister of economic affairs and finance. In the latter post he moved to nationalize the banks and to abolish payment of interest. His public following may have resulted from a combination of his religious background, his writings, his as-

sociation with the Ayatullah Khomeini at Paris, and his opposition to the domination of the government by the hard-line
clerics.

To return to the progress of establishing the republic, the
members of the National Consultative Assembly were elected
in the spring of 1980: religious figures were a large majority,
with one of their number chosen as Speaker. They voted to
change the name of the body to that of the Islamic Assembly.

On 4 November 1979, the Chancery, the office building of
the Embassy of the United States at Tehran, was taken by
storm and its diplomatic personnel made hostages. From that
date until the final release of fifty-two hostages on 20 January
1981, millions of words have been written on the subject.
Here a paragraph or two must suffice. As is well known, the
seizure of the personnel was related to the arrival on 22 October 1979 of the Shah at a New York City hospital. According to chargé d'affaires and hostage, Bruce Laingen, the first
demonstrations against the Shah's admission to the United
States occurred at the Embassy on 1 November and the government acted promptly to call off another scheduled for
later in the day and provided added protection. There was no
such protection on 4 November.

Day after day spokesmen for Iran justified the taking of
hostages. In November Bani Sadr stated: "Ever since the en-
try of the students into the U.S. Embassy—or more precisely
an important administration and espionage center in our world
—the propaganda machine has been conducting propaganda
every day and night intended to beguile you Americans and
the world into believing, that, under the Islamic regime, no
law or rule is respected, that traditions, which have for centuries been revered by humanity, are being undermined, that
your Embassy, which is regarded as part of your country's
territory, is stormed, that persons who enjoy diplomatic immunity are taken as hostages . . ." Demands for the return to
Iran of the Shah and his wealth became increasingly heated.
In December 1979 the Ayatullah Khomeini stated: "We want
you to surrender him to us. He is the same person who killed
our youth, who roasted our young people in burning pots, who
charred them on fire, and who cut their limbs."

Three areas of national life in Iran from the establishment of the Islamic Republic until the spring of 1981 will be examined: these are society, politics, and economics.

Society: What is the government of the Islamic Republic? Is it a theocracy, a state governed by clerics claiming divine sanction; a prelacy, a church government conducted by high-ranking clerics; or a monocracy, the rule by a single person? As long as the Ayatullah Khomeini remains active, it is a monocracy, as he has unlimited powers over the executive, legislative, and judicial branches. After his death it becomes, temporarily at least, a theocracy.

His rules of proper social behavior stem from prohibitions and admonitions contained in the Qoran and in the sayings of Muhammad, from traditional Muslim manners and customs, and from his own attitude toward perceived domestic and foreign enemies. Internal enemies are charged with "corruption on earth," and "warring with Allah and his emissaries." Hasty trials may result in immediate executions. On 8 May 1980 Mrs. Farrokhrou Parsa was executed after having been convicted of making education in Iran dependent upon "imperialist Western culture": she was a former minister of education, a former member of the Majlis, a physician, and an active member of the Women's Organization of Iran. At Kerman on 4 July 1980 two men and two women convicted of sexual offenses, clad in white robes, masked in "hoods of death," and buried up to their chests, were stoned to death, with the judge of the revolutionary court casting the first stone: in fifteen minutes they were dead.

In June 1981 legal codes covering certain crimes were replaced by a Law of Retaliation, approved by the prime minister and the cabinet and sent to the Assembly for ratification. Its numerous articles enables victims to take vengeance against offenders. Article 1. "Murder, if premeditated is subject to retaliation, and the parent, guardian or the next-of-kin of the victim may, subject to the approval of an Islamic judge or his representative, slay the murderer in accordance with the stipulation that will be stated." The stipulation is that the victim is innocent as regarded by Islamic law. Article 57. "Amputation of a limb, or an injury to it, if premeditated, is subject to retaliation. Accordingly, the parents, guardian or the next-of-kin of the victim may, subject to the approval of an Islamic judge or his representative, carry out the retaliation in keeping with the stipulations set by the Islamic judge." Article 63. "An injury which is inflicted upon a criminal by way of retaliation, must be equal in width and length to the injury suffered by the victim. The depth of the retaliatory wound must also be equal to that of the victim's wound. . . ."

Jews, Christians and Armenian Christians have, as promised by the Ayatullah Khomeini, suffered harassment and persecution. The Bahai sect which numbers between 300,000 and 500,000 adherents in Iran has been branded as heretical by the Islamic judges and its members deprived of all civil rights. A number have been executed charged with "treason to Islam," and "creating discord and disunity among Muslims."

Other internal enemies of the republic includes those who are alleged to have ties with the West. Khomeini identifies them as "lawyers, conspiring writers, journalists, corrupt intellectuals, secular liberals, human rights activists, democrats and critics of the regime." Overriding a Constitutional guarantee, Khomeini declared that the Kurdistan Democratic Party was illegal, thus foreshadowing the closing down of other political parties by less eminent authorities.

The harassment of internal enemies continued unabated. Early in June 1981 Amnesty International stated that at least 1,600 men and women had been executed.

The consumption of alcohol, pork, and caviar is forbidden. Narcotics, gambling, and usury are prohibited. Abortion is forbidden. Possible sexual offenses and immoral acts are monitored by the Bureau to Fight Forbidden Acts which has punitive powers. Principle 21 of the Constitution which concerns women is primarily focused on their role as mothers. In public they are to be enveloped in the chador; they are not to bathe in mixed company, although later Khomeini modified his insistence on the chador to accept "modest dress." Beauty shops are banned as "dens of moral corruption." The Ayatullah Khomeini has branded music as evil, adding that it is, "no different from opium . . . stupifies persons listening to it . . . dulls the mind because it involves pleasure and ecstasy . . . a betrayal of the nation and of the youth." When asked if he considered the music of Bach, Beethoven, and Verdi evil, he replied, "I do not know those names." In 1981 Iran did not change to daylight saving time as in the past, because it was now "contrary to the customs and character of Islam."

There are indications that efforts are being made to eradicate the pre-Islamic history and culture of Iran. Tens of millions of school books have been recalled, and new ones will place a "proper" emphasis on the Arab origin of Islam. As one example of steps being taken, those oil fields adjacent to islands in the Persian Gulf which had names coming from the Sasanian and other early dynasties have had those names replaced by ones related to Arab history and geography.

The Tehran press, or at least that segment of it which reaches the United States, fails to report on intellectual, cultural, and artistic life, and to list films shown, and it may be that all such activity has come to a standstill.

Politics: Political parties and associations proliferated before and after the monarchy was overthrown, but by mid-1981 most of them were no longer permitted to function. Such groups included the Committee for the Defense of Human Rights; the National Front of Iran which had been headed by Karim Sanjabi and Shapur Bakhtiar; the Party of the Iranian Nation, headed by Daryush Furuhar; and the National Democratic Party, led by Hedayatullah Matin-Daftari.

The Islamic Republican Party is *the* party of the republic. It was headed by the Ayatullah Muhammad Husayn Beheshti. The "pure" arm of the party is the Hezb-i Allah, the "Party of God," and its activist arm is the Fedayan-i Islam whose involvement in political assassinations is recorded in an earlier chapter.

In August 1979 there were rather widespread protests against new press regulations, which included among their restrictions any criticism of the Ayatullah Khomeini and the religious regime. The regime reacted with destructive attacks by the Pasdaran on the offices of the Tudeh Party, the Mujaheddin-i Khalq, and the Fedayan-i Khalq and these groups were forbidden, for a brief period, further activity. The Tudeh Party of Iran has a long history of legal and illegal activity, as detailed in Chapter VI. The Mujaheddin-i Khalq was a radical Islamic group which had been responsive to the teachings of Dr. Ali Shariati. By this time some of his teachings were unwelcome to the regime, such as his opposition to the extreme traditionalists and to his statement that "when it is said, 'Rule belongs to Allah,' the meaning is that rule belongs to the people, not to those who present themselves as the representatives or the sons of Allah. . . ." Before its suppression the group was alleged to be able to muster 5,000 guerrillas. The Fedayan-i Khalq was a radical leftist group: many of its estimated 3,000 armed men were trained by the P.L.O. outside of Iran.

A few political groups have survived. Most important among them is the Republican Party of the Muslim People, headed by the Ayatullah Shariatmadari, a respected figure with large popular followings in Khorasan and Azerbaijan who holds moderate views relating to the political activity of the clerics: he was harshly attacked in statements by the Ayatullah Beheshti and others within the regime. In the Islamic Assembly, Mehdi Bazargan headed a Freedom Movement which supported Bani Sadr and suffered the hostility of the Islamic Republican Party. Other parties may exist but are not in the news; an example may be the Ranjbari "Toilers" Party which claims to be active throughout the country and which has as its motto: "Not Russia, not Amrika, long live self-independent Iran." Ever since the appearance of the Constitution of 1906, periods of comparative public freedom have given rise to a proliferation of associations, and the present atmosphere is not as yet entirely hostile to the existence of new groupings in the political arena. Nor has public expression been entirely suppressed. In February 1981, some 133 university professors, journalists, novelists, poets, and others twice charged the regime with torturing political prisoners and of making "consistent and increasing attacks upon democratic rights and liberties." They demanded individual liberty, freedom of assembly, freedom of the press, and the opening of all universities (which were closed in the summer of 1980).

The war with Iraq was further destabilizing the political climate, with Iraq inciting the Arabs of Khuzistan, the Baluchis, and the Kurds to rise up against the regime. Actually, Kurdistan had already been in revolt against the government. *Economics*: In viewing the economic situation from 1978 into 1981, it is appropriate to quote from the Annual Report for 1357 (March 1978-March 1979) of the Bank Markazi Iran: "The year 1357 marked a turning point in the history of political and social strife in Iran. The persistent and widespread strikes in all sectors of the economy—and particularly those sustained by the most underprivileged and deprived class—resulted in an economic paralysis which played a decisive role in the downfall of the previous regime and the ultimate victory of the people's revolution."

The economic paralysis contained factors other than the widespread strikes. There was an alarming loss of skilled managers as entrepreneurs, bankers, industrialists, and other motivated professionals abandoned hospitals, factories, and burned-out banks. An abrupt halt to the construction industry threw 1 million men out of work. Many major projects were terminated until it was estimated that 3 million out of a work force of 10 million were unemployed. Projects terminated included the Tehran subway; the new Tehran airport; new highways; a number of nuclear power plants; a second natural gas pipeline to the USSR; the Shahestan Pahlavi municipal center at Tehran; a Peugot-Citroen and a Volkswagen assembly plant at Tehran; the Shahinshahr city, to the north of Isfahan; the Shah Reza Industrial Plant, to the south of Isfahan; and skyscrapers for the Industrial Credit Bank and the Foreign Trade Bank. It could be well argued that all these projects were foreign to the future of Iran as envisaged by the religious leaders of the Islamic Republic.

In 1980 inflation stood at 33 percent in the major towns and by 1981 it is believed to have increased to 50 percent: Many basic statistics are no longer available.

In the eighth edition of this work, it was stated that several domestic problems were certain to plague Iran in the future. They were: a shortage of skilled manpower, a shortage of housing, and shortages in agricultural production. While the demand for skilled manpower fell off with the cancellation of major projects, shortages of housing and of foodstuffs became more acute. In 1975 imports of foodstuffs were valued at $1.4 billion, in 1978 $3.6 billion, in 1979 $9 billion, and in 1981 an estimated $10 billion. These imports included wheat, rice, barley, sugar, tea, live sheep from Turkey and frozen lamb and mutton from Australia and New Zealand. In 1981 300,-000 tons of meat, 800,000 tons of sugar, and 1.2 million tons of wheat were to be imported. Rationing was introduced in November 1980. Kerosene and electricity were on the list, as were numerous items such as sugar, vegetable oils, and detergents which were sold at subsidized prices. Private cars were entitled to six gallons of gasoline a month.

There were numerous fiscal problems. In 1978 millions of Persians took their savings out of the banks and strenuous measures had to be taken to establish liquidity and prevent failure of the banking system. The circulation of money dropped and smuggling of currency was halted by the introduction of new currency. All banks were nationalized and merged into four groups: industrial, agricultural, construction, and commercial. Regulations of 21 March 1980 stated that banks could make a maximum 4 percent annual service charge on loans. A minimum amount, not stated, on savings accounts was to be augmented annually by a share of the profits of the banks from their investments.

In November 1980 private dealings in foreign currencies were prohibited and all such holdings were to be offered to banks for purchase. In 1976 the official rate of the rial was 67.75 to the U.S. dollar and foreign exchange was free of controls. In 1981 the official rate was 81, along with a "market rate" of 155, while the black-market rate was around 220 rials to the dollar.

For years the previous governments had labored to collect increasing sums from income taxes. Now such taxes were to be replaced by a voluntary offering of 20 percent of earnings to be handled by the clerics.

Some features of the economic situation seemed harmful to the bazaaris who had given such strong support to the Ayatullah Khomeini. Loaning money to wholesalers and importers had been a most active business, while now usury is prohibited. Also, religious figures have called for the elimination of the middlemen and for the nationalization of imports.

In September 1980 Iraq revoked a 1975 border agreement with Iran and presented several demands: Iraq to control the Shatt al-Arab; Iran to grant a measure of autonomy to the Arab population of Khuzistan; Iran to evacuate three small islands in the Persian Gulf near the Strait of Hormuz; and Iran to agree not to try to export its revolution to the Arab states. Iraq invaded Khuzistan and in fairly short order took the ports of Khorramshahr and Bandar Khomeini (formerly Bandar Shah), and beseiged Abadan, Ahwaz, and Dizful. Air action devastated the refineries and tanker terminals of both nations, and resulted in great damage to the major industrial plants of Khuzistan. Some two million people fled north to be sheltered and fed by the government. Iran's appeal to the International community to denounce the invader and to offer it help, found no sympathetic hearing because of Iran's earlier defiance of international bodies and governments over the retention of the American hostages. Iran's oil exports were cut off, with the exception of small amounts from the offshore fields down the Gulf. It was estimated that in 1980 total income from oil would drop to less than $18 billion. The crude output averaged 134 million barrels a day, with much of the total diverted to domestic consumption.

There may be an air of unreality about statements made concerning the economy of Iran. Bani Sadr was the author of *Eqtesadi Towhid*, "Economics of Unity," and of articles published in Paris: his views are anticapitalist and anti-imperialist and contain arguments that seem difficult to follow. However, in April 1981, faced with reality, Bani Sadr offered a concerned appraisal of the economy of Iran in acute difficulty. The GNP declined 9 percent in 1978, 13 percent in 1979, and not less than 10 percent in 1980. Money in circulation increased from $11 billion to $20 billion: "in view of the negative growth of the economy during the past three years, such a figure is very alarming." And, "from July until September 1980, $2.8 billion in bank deposits were withdrawn, resulting in rapid price increases." In the year ending in March 1979, "oil production decreased 29 percent with a further 23 percent for the following year." "What now determines the level of our oil production is home consumption, and we shall export it to the extent of meeting our foreign currency requirements, and not for meeting the budget deficit." "This," he said, "was one of the achievements of the revolution." Turning to agricultural production, after a rise of 6 percent in 1978, it declined 3.5 percent in 1979 and in 1981 the decline continued to be 3.5 percent.

Statements by Bani Sadr and other officials charge the previous regime, which had allocated tremendous sums to bringing vast areas under irrigation and to the purchase and manufacture of tractors and other machinery, with making Iran dependent for food supplies on other nations.

In August 1979 the then Minister of Agriculture stated his views: "Iran has enough cultivable land and useful water, [and] if we make correct use of all our human forces and all the natural potentials in our land and water throughout the country, we can provide food for a country like Pakistan. . . . They [the previous regime] spent fifteen years to ruin the country's agriculture. I think that if political difficulties do not arise, then in ten years we will reach the stage whereby we will become a net exporter of certain agricultural products."

In the spring of 1981 the Ayatullah Khomeini himself was saying that it would take as many as ten years for Iran to achieve economic independence. However, if the present state of society and of politics seems somewhat chaotic, the economic future appears disastrous. The disruption of the oil-based national economy and the prolongation of the conflict with Iraq are leading to a serious decline in living standards, marked by such features as rampant inflation, rationing of basic necessities, massive unemployment, and a growing black market, while the threatening shortages of foodstuffs raise the specter of starvation in Iran.

In the summer of 1981 dramatic events resulted in the strengthening of the hold of the monocracy over public activity and private life. President Bani Sadr came under harsh attacks by the hard-line clerics, and in early June the Ayatullah Khomeini abruptly discarded his role as moderator in these disputes and made repeated attacks on the president. On 7 June the prosecutor banned Engelab-i Islami, "Islamic Revolution," the daily paper of Bani Sadr, and on 10 June Khomeini dismissed the president as commander in chief of the armed forces. The following day Bani Sadr bitterly denounced the Islamic Republican Party and then vanished from sight. He surfaced in Paris at the end of July.

In Tehran supporters of Bani Sadr fought in the streets with the Hezb-i Allah and during successful impeachment proceedings against the president on 20 and 21 June more severe fighting occurred. On 22 June Khomeini dismissed Bani Sadr. On 29 June a bomb destroyed the headquarters of the Islamic Republican Party, killing the Ayatullah Beheshti and seventy-one others, including twenty members of the Islamic Assembly. Khomeini stated that the Mujaheddin-i Khalq, acting as "agents of the United States," were responsible and then called for a nation-wide purge of that organization. In a week some 140 people had been executed.

A Presidency Council had taken over the duties of the deposed Bani Sadr: Ayatullah Beheshti, Prime Minister Muhammad Raji and Speaker of the Islamic Assembly Hojatol Islami Hashemi Rafsanjani. Beheshti was replaced by Abdul Karim Musavi Ardabeli, pending the dissolution of the council on the election to the presidency of Rajai on 24 July.

Brief mention was made earlier of the Mujahhedin-i Khalq, "Crusaders of the Masses." The group early supported Khomeini, concealing its ties with the P.L.O., its relations with Moscow, its Marxist program and the communist background of its leaders, including Masud Rajavi. Rajavi accompanied Bani Sadr to Paris. The Fedayan-i-Khalq, "Devotees of the Masses," mentioned earlier, was another element of the radical left. The Tudeh Party remained quiescent, although active among the students, expressing allegiance to Khomeini and hoping to escape a purge. With the Islamic Republican Party supported by the Islamic Assembly, the revolutionary courts, the Pasdaran and the "club-wielders" in the street, the monocracy should be able to cope with any communist inspired challenge.

Early in July 1981, the Ayatullah Khomeini went into

seclusion for the fasting month of Ramadan. Living longer than had been expected by many observers, his choice as successor remained the Ayatullah Husayn Ali Montazeri. A former student of Khomeini, he was imprisoned as an opponent of the previous regime between 1974 and 1978. Named by Khomeini as the leading religious figure in Tehran, he conducts the Friday mass prayers, rifle in hand. He advocates the export of Iran's Islamic revolution to other countries and has condemned both liberal and leftist groups in Iran as having been "created and formed for American purposes."

Previous chapters describe the Persians fear and hatred of Great Britain and Russia and Iran's efforts to attract the United States as a powerful friend who would help support an independent Iran. Americans and American institutions came to be much admired, and many thousands of Persians sought an education in the United States. How then, did Amrika become the "Great Satan" of Khomeini? Khomeini convinced his followers that Amrika had supported the late Shah in its interests and against those of the nation. Also, Khomeini vehemently denounced the invasion of American "culture" which was alleged to have introduced rock music, blue jeans, alcohol, discotheques, mixed bathing, and pornography. If there is an answer as to how such an invasion may be halted and negated, it has not been found in other countries where similar invasions are deplored.

Not too many years ago the "hidden hand" of the British was believed to be responsible for the country's political ills. However, since the Islamic revolution Great Britain has managed to maintain a very low profile, not being attacked by the Ayatullah Khomeini. Reliable reports of daily events in Iran came through Reuters, a British press service, which was expelled in July 1981.

While the fear of hostile actions by the USSR has not lessened, it may be noted that Khomeini has been cautious in expressing such a view. He condemned the Soviet invasion of Afghanistan, but there is no record of direct Persian aid to the Afghans fighting against the Russians. Currently, improved economic relations in the fields of trade and import facilities are more important than political invective against the USSR.

In general, the Islamic Republic continues to reduce its contacts with foreign powers as part of moving toward its goal of total independence. As long as this trend continues there can be no possibility of renewed diplomatic relations with the United States.

As the overthrow of the monarchy was so unexpected, what caused the Persian to act with such sudden rage? How was it that Muslim clerics who had agitated for the liberal Constitution of 1906, could be followed by clerics who would destroy the public freedoms guaranteed in their Constitution of 1979? How was it that the masses who chose the monarchy over the charismatic figure of Dr. Mossadeq in August 1963, could discard the single stable institution which had held the country together for over two thousand years?

Social patterns and social behavior are discussed in Chapters III and V, and these paragraphs may be worth rereading. Negativism pervades Persian life: it finds expression in a lack of faith in public institutions; destructive criticism of groups, classes and individuals; and an unwillingness to engage in sustained group activity. Negativism was a force against the monarchy, but obviously not the only one. By the late Shah and by his father, Reza Shah, the Persians had been told that the country must be Westernized and that they must work to achieve the material progress of the West. This emphasis on a work ethic had never been a feature of Persian life, and it may be argued that these Persians did not want to be hard-working modern men, but wanted to be born again into the traditional society with its fewer pressures on the individual. Thus, they welcomed appeals to discard Westernization and its companion, modernization. Less specific than these two features was a background of conflict, rivalry, and instability within a society which lacked a basic homogeneity and that nurtured violent confrontation.

A vision of the future was contained in Bani Sadr's statement of 12 June 1981 to the Iranian people. It read, in part: "Economic independence is being destroyed. . . . Political independence will be lost without economic independence as well as through the existence of internal insecurity and foreign wars. . . . Permanent clashes will be added to the recent armed clashes. . . . The imposed war will wear out completely the Iranian and Iraqi forces if it continues any longer. . . . I tell you now that the dictatorship is not yet consolidated and if you do not resist it, tomorrow, linked with foreign hegemony, it will do what I said and even worse than that against you."

In the interim a vicious struggle continued between the dominant clerics and the Mujahhedin-i Khalq, of whom some 1,800 were executed by late 1981, while the clerics lost President Rajai and Prime Minister Muhammad Bahonar to their bombs. This struggle could exhaust both parties to the point where the military could intervene to maintain public order and to ally itself with the stronger survivor.

SELECTED BIBLIOGRAPHY

* Titles so marked are recommended to readers on the basis of their general interest and of their availability.

THE HISTORY AND CULTURE OF IRAN

A Catalogue of Qajar Paintings of the 18th and 19th Centuries, Tehran, 1971. Not paged.

Browne, E. G., *A Literary History of Persia,* Cambridge, 1928. 4 vols. 2180 pp. Reprinted.

Christensen, Arthur, *L'Iran sous les Sassanides,* Paris, 1944. 560 pp.

Culican, William, *The Medes and the Persians,* London, 1965. 260 pp.

Duchesne-Guillemin, Jacques, *La Religion de l'Iran Ancien,* Paris, 1962. 411 pp.

Edwards, C. A., *The Persian Carpet. Survey of the Rug Weaving Industry of Persia,* London, 1953. 384 pp. and 419 ill.

*Ghirshman, R., *Iran from the Earliest Times to the Islamic Conquest,* Penguin Books, 1954. 367 pp.

Ghirshman, R., *The Parthians and Sassanians,* London, 1962. 401 pp.

Godard, André, *The Art of Iran,* London, 1965. 358 pp.

Herzfeld, Ernst, *Iran in the Ancient East,* New York, 1941. 363 pp.

Hinz, Walther, *The Lost World of Elam,* London, 1972. 192 pp.

Huot, Jean-Louis, *Persia. I. From its Origins to the Achaemenids,* London, 1965. 219 pp.

Lockhart, Laurence. *The Fall of the Safavi Dynasty and the Afghan Occupation of Persia,* Cambridge, 1958. 584 pp.

*Lockhart, Laurence, *Persian Cities,* London, 1960. 188 pp.

Lukonin, Vladimir, *Persia. II.,* Geneva, 1967. 233 pp.

Olmstead, Albert T. E., *The History of the Persian Empire, Achaemenid Period,* Chicago, 1948. 576 pp.

Pope, Arthur U., ed., *A Survey of Persian Art from Prehistoric Times to the Present,* London and New York, 1938-39. 6 vols. 2817 pp. and 1482 plates. Reprinted in 12 volumes.

Pope, Arthur U., *Masterpieces of Persian Art,* New York, 1945. 155 plates.

*Pope, Arthur U., *Persian Architecture. The Triumph of Form and Color,* New York, 1965. 287 pp.

*Porada, Edith, *Ancient Iran. The Art of Pre-Islamic Times*, London, 1965. 269 pp.

Schmidt, Erich, *Persepolis I, II, III*, Chicago, various dates.

Sykes, Percy, *History of Persia*, London, 1930. 2 vols. 1179 pp.

The Cambridge History of Iran. Vol. I. The Land, 1968. 804 pp. *Vol. V. The Saljuq and Mongol Periods*, 1968. 778 pp.

Vanden Berghe, L., *Archéologie de l'Iran Ancien*, Leiden, 1959. 285 pp. and 173 plates.

Welch, Stuart C., *A King's Book of Kings. The Shah-Nameh of Shah Tahmasp*, New York, 1972. 199 pp. and with over 50 splendid color illustrations.

Wilber, Donald N., *The Architecture of Islamic Iran: The Il Khanid Period*, Princeton, 1955. 208 pp.

Wilber, Donald N., *Persian Gardens and Garden Pavilions*, Tokyo, 1962. 239 pp.

*Wilber, Donald N., *Persepolis. The Archaeology of Parsa, Seat of the Persian Kings*, New York, 1969. 120 pp.

Wilber, Donald N., *Four Hundred Forty-Six Kings of Iran*, Tehran, 1972. 189 pp.

Wulff, Hans, *The Traditional Crafts of Persia*, Cambridge, Massachusetts, 1966. 404 pp.

The Character of the Civilization of Earlier Iran

*Arberry, A. J., ed., *The Legacy of Persia*, Oxford, 1953. 421 pp.

*Arberry, A. J., *Classical Persian Literature*, London, 1958. 464 pp.

Arberry, A. J., *Immortal Rose: An Anthology of Persian Lyrics*, London, 1948. 174 pp.

Arberry, A. J., *Fifty Poems of Hafiz, Texts and Translations*, Cambridge, 2nd ed. 1953. 187 pp.

Arberry, A. J., *Persian Poems: An Anthology of Verse Translations*, London, 1954. 239 pp.

Bowen, J.C.E., *Poems from the Persian*, Oxford, 1948. 105 pp.

Darke, H., *The Book of Government, or Rules for Kings*, London, 1960. 259 pp. A translation of the *Siyasat nama* of Nizam al-Mulk.

Donaldson, Dwight M., *The Shi'ite Religion*, London, 1933. 393 pp.

Field, Henry, *Contributions to the Anthropology of Iran*, Chicago, 1939. 2 vols. 706 pp.

Frye, Richard N., *The Heritage of Persia*, London, 1962. 301 pp.

*Herbert, Sir Thomas, *Travels in Persia, 1627-1629. Abridged and edited by Sir William Foster*, London, 1928. 352 pp. (If obtainable, the original London edition of 1638 is preferable.)

Lambton, A.K.S., *Landlord and Peasant in Persia*, London, 1953. 459 pp.

*Levy, Reuben, *A Mirror for Princes. The Qabus Nama by Kai Ka'us ibn Iskandar, Prince of Gurgan*, London, 1951. 265 pp.

Levy, Reuben, *The Persian Language*, London, 1951. 125 pp.

Levy, Reuben, *The Tales of Marzuban*, New York, 1968. 254 pp.

Massé, Henri, *Persian Beliefs and Customs*, New Haven, 1954. 516 pp.

*Morier, James, *The Adventures of Hajji Baba of Isfahan*. (Many editions of this marvelous tale are available.)

Rehatsek, E., *The Gulistan or Rose Garden of Sa'di*, London, 1964. 265 pp.

Rypka, Jan et al., *History of Iranian Literature*, Dordrecht, Holland, 1968. 928 pp.

Rubaiyat of Omar Khayyam, the Astronomer-Poet of Persia; Rendered into English verse by Edward FitzGerald, London, 1859. Countless later editions are available, as well as translations by several scholars.

*Stevens, R., *The Land of the Great Sophy*, London, 1962. 307 pp.

Voyages du chevalier Chardin en Perse, et autres lieux d'Orient ..., Amsterdam, 1711. 3 vols. 987 pp. Later editions, some abridged, are available.

COUNTRY AND PEOPLE AS SEEN BY CONTEMPORARY OBSERVERS

Barth, Fredrik, *Principles of Social Organization in Southern Kurdistan*, Oslo, 1953. 146 pp.

Barth, Fredrik, *Nomads of South Persia. The Basseri Tribe of the Khamseh Confederacy*, London, 1961. 159 pp.

Blunt, Wilfred, *A Persian Spring*, London, 1957. 252 pp.

*Browne, E. G., *A Year amongst the Persians*, Cambridge, 1927. 650 pp.

Costa, A. and Lockhart, L., *Persia*, New York, 1958. 45 pp. and 110 plates.

Cronin, Vincent, *The Last Migration*, London, 1957. 343 pp.

SELECTED BIBLIOGRAPHY

Donaldson, Bess A., *The Wild Rue. A Study of Muhammadan Magic and Folklore in Iran*, London, 1938. 216 pp.

Filmer, Henry, *The Pageant of Persia*, Indianapolis, 1936. 422 pp.

Ishaque, Mohammad, *Modern Persian Poetry*, Calcutta, 1943. 225 pp.

Mazda, Maideh, *In A Persian Kitchen*, Tokyo, 1960. 175 pp.

Mehdevi, Anne S., *Persian Adventure*, New York, 1953. 320 pp.

Merritt-Hawkes, O. A., *Persia: Romance and Reality*, London, 1935. 322 pp.

Monteil, Vincent, *Iran*, Paris, 1957. 191 pp.

Monteil, Vincent, *Les Tribus du Fars et la Sédentarisation des Nomades*, Paris, 1966. 156 pp.

Najafi, Najmeh, *Persia is my Heart*, New York, 1957. 245 pp.

Najafi, Najmeh and Hinckley, H., *Reveille for a Persian Village*, New York, 1958. 273 pp.

*Smith, Anthony, *Blind White Fish in Persia*, London, 1953. 231 pp.

Stark, Freya, *The Valleys of the Assassins . . .* , London, 1934. 365 pp.

Suratgar, Olive, *I Sing in the Wilderness*, London, 1951, 222 pp.

Sykes, P. M. and Khan Bahadur Ahmad Din Khan, *The Glory of the Shia World. The Tale of a Pilgrimage*, London, 1910. 279 pp.

Ullens de Schotten, M. T., *Lords of the Mountains. Southern Persia and the Kashkai Tribe*, London, 1956. 128 pp.

MATERIAL ON CONTEMPORARY IRAN

Agabekov, G., *Ogpu, the Russian Secret Terror*, New York, 1931. 277 pp.

Amuzegar, Jahangir and M. A. Fekrat, *Iran. Economic Development under Dualistic Conditions*, Chicago, 1971.

Arasteh, R., *Education and Social Awakening in Iran, 1850-1960*, Leiden, 1962. 144 pp.

*Arfa, General Hassan, *Under Five Shahs*, London, 1964. 457 pp.

Avery, Peter, *Modern Iran*, London, 1965. 527 pp.

Baldwin, George, *Planning and Development in Iran*, Baltimore, 1967.

SELECTED BIBLIOGRAPHY

Balfour, Hon. J. M., *Recent Happenings in Persia*, London, 1922. 307 pp.

*Banani, Amin. *The Modernization of Iran, 1921-1941*, Stanford, 1961. 191 pp.

Bharier, Julian, *Economic Development in Iran 1900-1970*, London, 1971.

Binder, Leonard, *Iran. Political Development in a Changing Society*, Berkeley, 1962. 362 pp.

Denman, D. R., *The King's Vista: A Land Reform Which Has Changed the Face of Persia*, London, 1973. 368 pp.

Eagleton, William, Jr., *The Kurdish Republic of 1946*, London, 1963. 142 pp.

Elwell-Sutton, L. P., *Persian Oil: A Study in Power Politics*, London, 1955. 343 pp.

Fatemi, N. S., *Diplomatic History of Persia, 1917-1923; Anglo-Russian Power Politics in Iran*, New York, 1952. 331 pp.

Fatemi, N. S., *Oil Diplomacy. Powder Keg in Iran*, New York, 1954. 405 pp.

Ford, Alan W., *The Anglo-Iranian Oil Dispute of 1951-1952*, Berkeley, 1954. 348 pp.

Frye, Richard N., *Iran*, London, 1960. 126 pp.

Haas, William S., *Iran*, New York, 1946. 273 pp.

Hamzavi, A. H., *Persia and the Powers*, London, 1946, 125 pp.

Lambton, A.K.S., *The Persian Land Reform 1962-1966*, Oxford, 1969. 386 pp.

*Lenczowski, George, *Russia and the West in Iran, 1918-1948*, Ithaca, 1949. 383 pp.

Millspaugh, Arthur C., *The American Task in Persia*, New York, 1925. 322 pp.

*Millspaugh, Arthur C., *Americans in Persia*, Washington, 1946. 293 pp.

*Mohammad Reza Shah Pahlavi, *Mission for My Country*, London and New York, 1961. 336 pp.

Mohammad Reza Shah Pahlavi, *The White Revolution*, Tehran, 1967. 177 pp.

Motter, T. H. Vail, *The Persian Corridor and Aid to Russia*, Washington, 1952. 545 pp.

Ramazani, Rouhollah K., *The Foreign Policy of Iran. 1500-1941. A Developing Nation in World Affairs*, Charlottesville, 1966. 330 pp.

Shuster, W. Morgan, *The Strangling of Persia*, New York, 1912. 423 pp.
Skrine, C., *World War in Iran*, London, 1962. 330 pp.
*Upton, Joseph M., *The History of Modern Iran: An Interpretation*, Cambridge, 1960. 163 pp.
Vreeland, H. H., ed., *Iran*, New Haven, 1957. 429 pp.
Warne, William E., *Mission for Peace: Point Four in Iran*, New York, 1956. 320 pp.
Weaver, Paul E., *Soviet Strategy in Iran 1941-1957*, Washington, 1958. 277 pp.
Wilber, Donald N., *Contemporary Iran*, New York, 1963.
*Yar-Shater, E., ed., *Iran Faces the Seventies*, New York, 1971.
Yeselson, Abrahim, *U.S.-Persian Diplomatic Relations, 1883-1921*, New Brunswick, 1956. 252 pp.
Zabih, Sepehr, *The Communist Movement in Iran*, Berkeley, 1966. 279 pp.

RISE OF THE ISLAMIC REPUBLIC

Akhavi, Shahrough, *Religion and Politics in Contemporary Iran*, Albany, 1980.
Fischer, Michael M., *Iran: From Religious Dispute to Revolution*, Cambridge, Mass., 1980.
Graham, Richard, *Iran: The Illusion of Power*, London, 1979.
Halliday, Fred, *Iran: Dictatorship and Development*, New York, 1979.
Hoveya, Feridun, *The Fall of the Shah*, New York, 1980.
Keddie, Nikki, *Iran: Religion, Politics and Society*, London, 1980.
Ledeen, Michael and William Lewis, *The American Failure in Iran*, New York, 1981.
Muhammad Reza Pahlavi, *Answer to History*, New York, 1980.
Nobari, Ali Reza, ed., *Iran Erupts*, Stanford, 1978.
Rubin, Barry, *Paved with Good Intentions: The American Experience in Iran*, New York, 1980.
Saikal, Amin, *The Rise and Fall of the Shah*, Princeton, 1980.
Shariti, Ali, *On the Sociology of Islam*, Berkeley, 1979.

INDEX

Abadan Island, refinery at, 268;
airfield at, 331; port, 331; riot
at, 141
Abadeh, town, 181
Abaqa, Il Khan ruler, 49, 50
'Abbas, Shah, Safavid ruler, 61,
63, 77, 78, 114
'Abbas II, Safavid ruler, 63
'Abbasid Caliphate, 39, 82
Ab-i-Diz River, 293
Abu 'Ali ibn Sina (Avicenna),
philosopher, 40, 93
Abu Bakr, Caliph, 37, 81
Abu Bakr, Salghurid ruler, 53
Abu Mansur Faramuz, Kakuyid
ruler, 41
Abu Mansur Sebuktigin, Ghaznavid
ruler, 44
Abu Sa'id, Il Khan ruler, 52, 53,
110
Abu Sa'id, Timurid ruler, 57
Achaemenes, 26
Achaemenid Empire, 27ff, establish-
ment, 27; military campaigns, 27,
28; army, 28; roads, 28, 29;
languages, 29; religion, 80;
architecture, 95ff; art, 98
'Adud al-Dawla, Buyid ruler, 41,
77
Afghanistan, Helmand River in,
252; war with, 67; relations
with, 252
Afsharid dynasty, 64, 65
Agha Muhammad Khan, Qajar
ruler, 65, 66
Agricultural Bank, 175, 176
agriculture, 284ff; machinery for,
297
Ahmad, Buyid ruler, 40, 41
Ahmad, Il Khan ruler, 50
Ahmad Shah, Qajar ruler, 71, 126
Ahuramazda, deity, 79, 80, 102
Ahwaz, town, 10, 265
air force, 255
airports, 330
air services, 330
ajami, foreigners, 35

akhlaq, ethics, 84, 94
'Ala' al-Dawla Muhammad,
Kakuyid ruler, 41
'Ala' al-Din, Khwarazm-Shah
ruler, 48
'Ala, Husayn, 141, 151
al-Biruni, author, 40, 88, 110
Alam, Asadullah, 155, 233
Alamut, fortress, 47, 49
Alborz College, 203
Alborz Range, 5
Alexander the Great, 299; Persep-
olis burned by, 30; campaigns of,
30, 31; marries Roxana, 30;
at tomb of Cyrus, 31; plans a
new world state, 31; marries
daughter of Darius III, 31;
death of, 31
al-Ghazali, philosopher, 89, 94
'Ali, Shi'a Caliph, 60, 81, 82, 109
'Ali ibn Buya, ruler, 40
'Ali Shir Nawa'i, poet, 92, 113
Allies, in World War I, 71, 72; in
World War II, 132, 133, 134
Alp Arslan, Seljuq ruler, 45
Alptegin, Ghaznavid, 43, 44
aluminum, 265
American Legation, founded, 68
American missions, financial, 318;
military, 251; to Security Guard,
251
American schools, 202, 203
Amini, Dr. 'Ali, 150, 154, 155, 175
Amiranian Oil Company, 268
'Amr ibn Layth, Saffarid ruler, 42,
43
Amu Darya River, 44, 46, 89
Anahita, goddess, 33, 80, 102
Anarak, mines, 266
ancient mounds, 19, 20
Anglo-Iranian Oil Company, 142,
143, 149; terms of concessions,
268; production, 268; employees,
268; nationalization of, 141,
269; negotiations with, 149, 269;
member of the consortium, 150,
269, 270; becomes British Petroleum Company, 150, 269

horses, 298, 299; Alexander the Great inspects the Nisaean herd, 299

hospitals, 212; American, 212; other foreign, 212

Hotu cave, 17

Houses of Justice, 155

houses, village, 167, 168

housing, 314, 315

Hoveida, Amir 'Abbas, 156, 157, 249

Hulagu, Il Khan ruler, 48, 49, 51, 107

hunting, 300, 301

Husayn, son of 'Ali, 81, 82, 109

Husayn Bayqara, Timurid ruler, 58, 113

Ibn Khaldun, 86

Ibn Sina, *see* Abu 'Ali ibn Sina

ice houses, 186

Il Khans, Mongol rulers, 49f

Imam Reza, *see* Reza, Imam

Imams, 83

Imamzadehs, local shrines, 83

Imperial Bank of (Persia) Iran, 68, 69

Imperial Investigation Commission, 153, 247

Imperial Mosque, Isfahan, 114, 195

Imperial Organization for Social Services, 173

Imperial Square, Isfahan, 114, 198

imports, 316, 317, 318

industry, 302ff; film, 219, 220; heavy, 305; pharmaceutical, 307

Information, Ministry of, 213

Inshushinak, god, 23

investment, foreign, 308, 309

Iran, geographical position of, 3; trade routes across, 4; area, 4; frontiers of, 4; topography of, 4, 5, 8, 9, 10; drainage basins, 9, 10; rivers of, 10, 11; desert regions, 12, 13; rainfall of, 13; snow of, 13, 14; seasonal changes, 14; winter, 14, 15; prehistoric period, 16-22; pre-Achaemenid period, 22; arrival of Aryans in, 25; groups of Iranians, 25, 26; conquest by Alexander the Great, 30, 31; in Seleucid period, 31,

32; in Parthian period, 32, 33; in Sasanian period, 33-37; Arab conquest of, 37-39; early Islamic period, 39-44; in Seljuq period, 44-47; in Mongol (Il Khanid) period, 47-55; in Timurid period, 55-58; in Safavid period, 60-64; in Afshar, Zand and Qajar periods, 64-73; character of culture and society, 74-93; in Pahlavi period, 125-153; foreign relations of, 156f; mineral resources of, 263ff; petroleum in, 267ff; cultivated areas of, 284; agricultural products, 284ff; forests of, 287, 288; fruits of, 284, 285; use of water in, 289ff; farming in, 166ff; animals of, 298ff; birds of, 299, 300; economic development of, 259ff; industry in, 302ff; foreign trade of, 315ff; rug weaving in, 115, 116, 309, 310, 311; trade routes and roads of, 324ff; railways of, 326ff; aviation in, 330; government of, 229ff; population of, 160; population groups, 161; languages spoken in, 161, 163, 213, 214; nomadic tribes of, 163-165; village life, 166ff; towns of, 179ff; religion of, 79-84; flag of, 332; calendar of, 332; currency of, 333; weights and measures used in, 334

Iran Air, 330

Iran-America Society, 208

Iran-diberbad, head administrator, 35

Iran novin party, 155, 156, 229, 230, 231, 233

Iran party, 143, 231, 232

Iran-spahbad, military leader, 35

Iranian Carpet Company, 311

Iranian Oil Exploration and Producing Company, 270, 278

Iranian Oil Participants, 270

Iranian Oil Refining Company, 270, 278

Iranian Oil Services, 278

Iranian party, 233

Iranians, people, 25

railways, 326ff; *see also* Trans-
Iranian Railway
rainfall, 13
Rashid al-Din Fazl Allah, vazir and
author, 52, 60, 91
Ray, town, 193; pottery from, 106,
110; tomb at, 106
Razmara, General 'Ali, 140, 141
RCD, 250
Red Lion and Sun Society, 212, 224
relations, foreign, 250, 251, 252,
253, 254
religion, Muslim, 37, 38, 81f, 102,
162, 229
Religious Corps, 178
Resget, tomb tower, 104
Reuter, Baron, concessions granted,
67, 68
Reza, Cyrus 'Ali, crown prince,
154, 156, 246
Reza-i-'Abbasi, painter, 117
Reza, Imam 'Ali, shrine at Mashhad,
63, 83
Reza Shah Pahlavi, birthplace, 125;
military career, 125; marches on
Tehran, 125, 126; as Minister
of War and Prime Minister, 126;
named Shah, 126; aim to awake
Iran from lethargy, 126; policy
toward Great Britain, 127; attitude toward Russia, 127; creates
army, 127, 128; carries out social
changes, 127, 128; moves against
the clergy, 128; visits Turkey,
129, 130; orders state economic
controls, 303, 304; and World
War II, 131; fails to heed British
and Russian warnings, 132; abdi-
cates, 133; death, 133; mausoleum of, 193
Rezaieh Lake, *see* Lake Rezaieh
rial, currency, 333
rice, 284; dishes with, 184
roads, early, 28, 29, 322, 323; early
vehicular, 324, 325; modern,
325ff
Romanus Diogenes, Byzantine ruler,
45
Roxana, or Roshanak, wife of Alexander the Great, 30
"Royal Road," 29, 322
Rudaki, poet, 88

rug weaving, in Safavid period, 115,
116, 309; modern, 309, 310, 311
Russia, annexes Georgia, 66; war
with Iran ended by Treaty of
Gulistan, 66; war with Iran
ended by Treaty of Turkoman
Chai, 66; interests conflict with
those of Great Britain, 66, 67;
expansion in Asia, 68; fosters
Cossack Brigade, 68; founds
Discount Bank of Persia, 68;
gains Caspian Sea fishing rights,
68; loans made to Iran, 69;
builds carriage roads, 69; joins
Great Britain in 1907 agreement,
70; opposes Constitutionalists, 71;
demands dismissal of Shustar, 71;
during World War I, 71, 72;
terms of 1921 Treaty of Friendship, 72; commercial and economic relations with, 127; attitude of Reza Shah toward, 127;
forces enter Iran, 132; joins in
Tripartite Treaty, 134; military
supplies move through Iran to,
134, 135; requests oil concession,
135; halts Persian troops, 136;
alleged interference brought before Security Council, 136, 137;
fails to withdraw troops, 136;
proposes joint oil company, 137;
oil company rejected by Iran,
139; more recent desire for good
relations, 152, joint projects with
Iran, 251, 264; natural gas from
Iran to, 279
Rustam ibn Marzban Dushmanziyar,
Kakuyid ruler, 41

Sa'adabad Pact, 249, 250
Sa'di, poet, 53, 90; tomb of, 199,
200
Safavid period, 60ff; 113
Saffarid dynasty, 42, 43
Safi, Shah, Safavid ruler, 63
Sahand, Mount, 8
Saka, tribe, 32
salak, 210
Salamis, Battle of, 28
Salghurid dynasty, 53
Samanid dynasty, 43
Samarqand, Timurid capital, 111

Turan, home of Turic tribes, 89
Turbat-i-Shaykh Jam, shrine at, 111
Turki, language, 161, 162, 163
Turks, *see* Ottoman Turks. *See also*
Ghuzz Turks; Seljuq period
turquoise mines, 266
typhoid fever, 211
typhus, 210

'Ubaid-i-Zakani, writer, 54
Ulugh Beg, Timurid ruler, 57, 58, 111
'Umar, Caliph, 81
'Umar Khayyam, 90; tomb of, 115
Umayyad Caliphs, 38, 39, 82
United Kingdom, *see* Great Britain
United Nations, Iran joins, 135;
Iran appeals to, 136, 137
United States, opens Legation, 68;
participates in Teheran Confer-
ence and Tehran Declaration,
135; army in Iran during World
War II, 134, 135; alliances with,
152; relations of with Iran, 251;
missions to Iran, 250, 251, 318,
319; trade with Iran, 317, 318
University of Tehran, 150, 205, 206
'Unsur al-Ma'ali Kay Ka'us ibn
Iskandar, 40, 94
'Unsuri, poet, 88
Untash Gal, king, 23
uranium, 266
Urartu, kingdom of, 23-26
'Uthman, Caliph, 38, 81, 82
Uvakhshtra, 26
Uways, 55
Uzbek Turks, 61, 62
Uzun Khan, Aq-Qoyunlu ruler, 59, 60

Valerian, Roman ruler, 32, 34
vazir, official, 43
vegetable oils, processed, 289
vegetables, 284
venereal diseases, 210

village life, 120, 166ff, 167, 169, 171
villages, 160, 166ff
voting, 230, 231

wages, daily, 314
Wassmuss, German agent, 72
water supply, 289ff
weights and measures, 334
wells, power drilled, 292
wheat, 17; production of, 284
White Revolution, 155, 247
wine, 285
wool, 299; textiles, 306
World War I, Iran during, 71, 72
World War II, Iran during, 131-137

Xerxes, Achaemenid ruler, 28;
buildings at Persepolis, 96, 97

Ya'qub, Aq-Qoyunlu ruler, 59
Ya'qub ibn Layth, Saffarid ruler, 42
Yazdagird III, Sasanian ruler, 35

Zafar nama, history, 92, 113
Zagros Range, 5, 25
Zahedi, General Fazlollah, prime
minister, 146-151
Zaid Shi'ites, 83
Zand dynasty, 65
Zangi, Salghurid ruler, 53
Zarathushtra, 79, 80, 88
Zavara, mosque at, 105
Zayandeh River, 10, 194, 197; irri-
gation project, 293
zimmis, infidels, 37
Ziwiye, ancient site, 24
Ziyarid dynasty, 39, 40, 103
Zodiac, sign of the, 332
Zoroaster, *see* Zarathushtra
Zoroastrianism, *see* Mazdaism
Zoroastrians, modern, 162
Zur khaneh, house of strength, 119, 188

Index to "From Monarchy to Islamic Republic"